Foundations of Chemistry

Applying POGIL Principles

Fourth Edition

David M. Hanson

Stony Brook University - SUNY

Foundations of Chemistry
Applying POGIL Principles
Fourth Edition

by David M. Hanson
Department of Chemistry
Stony Brook University - SUNY
Stony Brook, NY 11794-3400
David.Hanson@StonyBrook.edu

Pacific Crest
906 Lacey Avenue, Suite 206
Lisle, IL 60532
630-737-1067
www.pcrest.com

ISBN: 978-1-60263-504-3

Acknowledgments

- Dan Apple, founder and president of Pacific Crest, is acknowledged as the motivating force and inspiration behind process-oriented education. His insights on activity design, classroom facilitation, and changing pedagogy are much appreciated.
- This work could not have been sustained without the encouragement of my wife, Colleen, and my colleague, Troy Wolfskill. Both provided inspiration, constant encouragement and help, and insights from their experience in teaching. Colleen was especially pivotal in convincing me that learning teams work and that it's process not content that should be the focus of education.
- The quality of the Fourth Edition was enhanced significantly through the fantastic skills of Denna Hintze-Yates (Pacific Crest, editing and production) and Heather Nehring (Pacific Crest, proofing); and the eagle-eye proofreading of David Moss (Moses Brown School) and Kathleen Cornely (Providence College), who also completely revised the *Instructor's Edition*. The suggestions received from numerous faculty, who adopted *Foundations of Chemistry* for use in their General Chemistry classes, were equally important in correcting typos and expanding the content to comprehensively cover two semesters of General Chemistry. Vicky Minderhout and her colleagues at Seattle University merit special mention in this regard, and also for modifying and contributing several activities based on their classroom experiences.
- Support from the National Science Foundation made it possible to develop, test, and revise activities for general and physical chemistry; share them with others; help others move from lecturing to more student-centered teaching strategies; and work toward enhancing process-oriented guided-inquiry activities with computer and web technology. The following grants supported this and related projects: DUE-9752570, DUE-9950612, DUE-0127650, DUE-0127291, DUE-0231120, and DUE-0341485.

Table of Contents

To the Instructor and Student

Process-oriented guided-inquiry learning (POGIL) is both a philosophy of and strategy for teaching and learning. It is a philosophy because its practice is based on specific ideas about the nature of the learning process as well as the expected outcomes of learning. It is a strategy because it provides a specific structure for teaching that is consistent with the way people learn, thus leading to the desired outcomes.

The goal of POGIL is to engage students in the learning process, helping them master the material through conceptual understanding (rather than memorizing and pattern matching), as they work to develop essential learning skills. Important skill areas for success in chemistry courses, college, and beyond include: information processing, critical and analytical thinking, problem solving, oral and written communication, teamwork, and metacognition (reflection on learning, self-assessment, and self-management).

To support this goal, POGIL activities utilize a learning cycle design that consists of exploration, concept formation, and subsequent application.

Each activity begins with an *Orientation* that sets the stage for learning. The importance of the activity is described in a *Why* statement. *Learning Objectives* and *Success Criteria* are identified along with prerequisite activities. The learning objectives describe what the student is expected to learn through completing the activity. The success criteria specify the measurable outcomes of this learning; they describe how the learner should be able to demonstrate that he or she has successfully learned the activity content. It is quick and easy to make up examination questions simply by looking at the success criteria for each activity.

Students then explore a *Model* in response to *Key Questions*. The model is any representation of what is to be learned. Through working to answer the Key Questions, students unlock the information present in the model, revealing its significance. These questions guide the learner in discovering the relevant concepts and in developing an understanding of them. *Information* is provided at the beginning and throughout an activity, but only when a *need-to-know* has been created.

The new knowledge is then applied in simple *Exercises* that build confidence, and then to higher-level applications, called *Problems*, that require learners to synthesize ideas, transfer their learning to new contexts, and demonstrate their problem-solving skills.

While students can complete an activity individually, the activities are most effective when used by students working in learning teams with an instructor acting as a coach, guide, or facilitator.

Many resources are available to help instructors teach in this new student-centered environment. They can be found at the Pacific Crest (www.pcrest.com) and POGIL (www.pogil.org) web sites. Both Pacific Crest and the NSF-supported POGIL Project sponsor workshops for faculty to introduce them to process-oriented guided-inquiry learning, assist them in developing facilitation skills, and guide them to materials for use in their courses.

To the Student

Changes in society, technology, and the world economy are occurring at increasingly faster rates. As a college graduate, you will need to be a quick learner, critical thinker, and successful problem solver to succeed in this rapidly changing environment. You will also need to be computer literate and demonstrate a high degree of skill in the areas of communication, teamwork, management, and assessment. This book is intended for use in courses where faculty are responding to these pressures and needs by making changes in the way they teach and using curricula that will actively engage you in learning, helping you to develop these essential skills.

This book is not like textbooks that you have used before. It does not provide you with information to read, memorize, and repeat during exams. Rather, it provides some representation, or model, of what is to be learned. Key questions guide your exploration of the model and help you unlock the information it contains, revealing its significance. The key questions help you discover the relevant concepts and develop an understanding of them.

Exercises then give you practice at applying these concepts in a straightforward context. They will help to build your confidence as you practice using your new knowledge to solve problems and answer questions. A few problems are included in each activity and your instructor may assign others. Problems are not as straightforward as the exercises; they often require the use of two or more concepts and the application of your knowledge in new contexts.

This book has been very carefully designed with the goal of helping you learn how to process information, analyze situations by asking yourself key questions, construct your understanding of chemistry, and develop the problem solving skills that you need to be successful in this course, in college, and beyond.

You will learn the most and have the most fun if you work on these activities with other students. Discussions among members of your learning team will produce different perspectives regarding the concepts and their use in solving problems. These discussions will help you to identify and correct misconceptions you may have and strengthen and deepen your understanding of chemistry.

You should use your textbook to resolve disagreements, to find answers to questions that arise, and to see examples of problem solutions. But it is through your understanding of the concepts and how to use them that you will be able to successfully answer exam questions and solve real-world problems. When you are working in a learning team, you should have two objectives: to understand the material yourself and to ensure that every other member of the team understands the material as well. Explaining ideas and helping others learn are among the best ways for you to deepen your own understanding and knowledge, giving you valuable insight (and skills) that will further ensure your success on exams, as well as when you're faced with real-world problem that must be solved.

We have found that this approach works for most students; they do better on exams, develop a deeper understanding of chemistry, recognize that they have become stronger learners, and have more fun along the way.

Activity 01-1

Units of Measurement

Why?

Units identify the scale that is used in making a measurement and are essential for the measurement to be meaningful. For example, if someone tells you that a person is 60 tall, you do not know whether they are referring to a child or possibly an adult until you know the units. The units could be inches or centimeters, but probably are not feet or meters. In your study of chemistry and its applications, you need to be familiar with the basic units that are used for mass, length, time, temperature, electrical current, and amount of substance. Prefixes to the units make it quicker and easier to write very large or very small numbers, e.g., 5,600,000 g = 5.6 Mg or 0.000001 s = 1 μs. Other units are derived from these basic units. For example, units of volume are derived from units of length, and units of energy are derived from units of mass, length, and time. You also need to be able to convert from one set of units to another because different countries, disciplines, and even sub-disciplines of chemistry often use different units for the same quantity. For example, in the United States speed limits are given as miles per hour; in many other countries speed limits are given as kilometers per hour.

Learning Objectives

- Identify the units used to measure physical quantities
- Become familiar with the prefixes used for larger and smaller quantities
- Master the use of unit conversion in solving problems

Success Criteria

- Associate units with physical quantities
- Replace prefixes by multiplying by appropriate numerical factors
- Identify, set up, compute, and validate unit conversions

Prerequisite

- Exponential notation

Model 1: International System of Units (SI Units)

Table 1

Physical Quantity	Name of Unit	Abbreviation
mass	kilogram	kg
length	meter	m
time	second	s
temperature	Kelvin	K
electrical current	ampere	A
amount of substance	mole	mol

(continued on the following page)

Model 1: International System of Units (SI Units), continued

Table 2

Prefix	Abbreviation	Meaning
tera	T	10^{12}
giga	G	10^{9}
mega	M	10^{6}
kilo	k	10^{3}
centi	c	10^{-2}
milli	m	10^{-3}
micro	μ	10^{-6}
nano	n	10^{-9}
pico	p	10^{-12}
femto	f	10^{-15}

Examples

Mass: A quarter-pound hamburger has a mass of 0.11 kg. 0.11 kg = 110 g

Length: A tall basketball player (7 feet) has a height of 2.15 m. 2.15 m = 215 cm = 2150 mm

Time: There are 3300 s in a 55 minute chemistry lecture. 3300 s = 3.300 ks

Temperature: Water freezes at 273 K.

1×10^3 g = 1000 g = 1 kg

1 cm = 0.01 m = 1×10^{-2} m

10 cm × 10 cm × 10 cm = 1000 cm^3 = 1 L (Note: L = liter)

1 ps = 1×10^{-12} s

Key Questions

1. What are the 6 basic units and their abbreviations listed in **Model 1** (Table 1)? (You need to have these units and their abbreviations memorized. Writing them without looking at the table will help you remember.)

2. What are the 10 prefixes listed in **Model 1** (Table 2), along with their abbreviations and meanings? (You need to have these prefixes, their abbreviations, and meanings memorized, and writing them without looking at the table will help you remember.)

3. How is the unit of volume (liter) derived from the unit of length (cm) in one of the Examples below **Model 1**?

$$1 cm^3 = 1 mL$$

4. A driver in Canada claims she was only doing 90 when the police officer stopped her. What else do you need to know to decide whether she was guilty of speeding?

miles per hour or kilometers per hour

EXERCISES

1. Write the number of seconds in a day (86,400 s) in exponential notation.

$$8.64 \times 10^4$$

2. Write the abbreviation for the units used when the number of seconds in an hour is expressed as 3.6.

s/hr 3600 3.6×10^3 $3.6 ks$

3. A laser pulse is 0.15 ns long. Write this time using exponential notation with units of s.

0.15×10^{-9} $.15 ns$ $\frac{10^{-9} s}{1 ns}$ $1.5 \times 10^{-10} s$

4. Michael Oher, offensive tackle for the Baltimore Ravens and subject of the book and movie *The Blind Side*, has a mass of 141,000 g. Write this mass using units of kg.

$141000 g$ $\frac{1 kg}{1000 g}$ $141 kg$

5. The mass of a car is about 1.4 Mg. Write this mass using exponential notation with units of g.

$1.4 Mg$ $\frac{10^6 g}{1 Mg} = 1.4 \times 10^6 g$

6. The diameter of a helium atom is about 30 pm. Write this length using exponential notation with units of m.

$30 pm$ $\frac{10^{-12} m}{1 pm} = 30 \times 10^{-12} m$

7. The diameter of the Earth is about 12,800 km. Write the diameter of the Earth using units of Gm.

$12800 km$ $\frac{10^3 m}{1 km}$ $\frac{1 Gm}{10^9 m} = 1.28 \times 10^{-2} Gm$

8. Determine how many ps there are in 1 s.

$1 \times 10^{-12} ps$

MODEL 2: A GENERAL CHEMISTRY PROBLEM-SOLVING STRATEGY APPLIED TO CONVERTING UNITS

Following this six step problem-solving strategy will help you solve more complicated problems. It is illustrated here for the case of unit conversion.

In a unit conversion problem, you need an *equality statement* that tells you how the two units are related, e.g. 1 quarter = 25 pennies or 2.2 lb = 1.0 kg.

From the equality statement, you can produce *conversion factors* that are used to convert one unit into the other, e.g. 1quarter/25 pennies or 1.0 kg/2.2 lb.

Example Unit Conversion Problem: Your pharmacist friend calls you and needs help to convert the mass of a medicine from ounces to drams. There are 5.00 ounces of the medicine; what is the mass in drams?

Strategy	Example
Step 1: Identify and record what is known or given.	**Known:** have 5.00 ounces of medicine
Step 2: Identify and record what is unknown and needs to be found.	**Unknown:** mass of medicine in drams
Step 3: Identify and record the concepts that connect what needs to be found to what is known.	**Connections:** look up an equality statement: 1 ounce = 15.75 drams conversion factors: 1 ounce/15.75 drams = 1 15.75 drams/1 ounce = 1
Step 4: Set up the solution using the connections.	**Setup:** apply the appropriate conversion factor 5.00 ounces × (15.75 drams/1 ounce)
Step 5: Do the mathematics to obtain the result.	**Result:** 78.8 drams, note the units of ounces cancel
Step 6: Check or validate your answer.	**Validation:** units correct, value is reasonable (~5 × 16)

KEY QUESTIONS

5. What units need to be converted in **Model 2**?

ounces to drams

6. How many conversion factors result from a single equality statement in the model? Explain.

1 because you only need to convert once

7. When the conversion factor was applied in Step 4 to obtain the result in Step 5 of the strategy, what happened to the ounces?

canceled out

8. In general, how can you identify whether you have used the correct unit conversion factor?

if units cancel out

9. Is multiplying by a unit conversion factor the same as or different from multiplying by 1? Explain.

yes because they equal eachother

EXERCISES

9. Michael Oher has a mass of 141 kg, see Exercise 4. What is his mass in lbs? (1.00 kg = 2.20 lbs)

141 kg $\frac{2.20 \text{ lbs}}{1 \text{ kg}}$ = 310.2 lbs

10. Was the lady from Canada in *Key Question 4* speeding if she was going 90 km/hr and the speed limit is 60 mi/hr? Explain. (1.0 mi = 1.6 km)

speed limit

60 mi/h $\frac{1.6 \text{ km}}{1 \text{ m}}$ = 96 km/h

11. Complete the following table.

Equality Statement	Conversion	Conversion Factor	Result
2.54 cm = 1.00 in	63 in to cm	63 in $\frac{2.54 \text{ cm}}{1 \text{ in}}$	
1.00 L = 1.06 qt	4.0 qt to L	4.0 qt $\frac{1 \text{ L}}{1.06 \text{ qt}}$	
36.0 pasos = 136.3 paras	95 paras to pasos	95 paras $\frac{36 \text{ pasos}}{136.3 \text{ paras}}$	

PROBLEMS

1. You are the purchasing agent at a start-up biotechnology firm. If sucrose costs $1.20 per pound, and a bottle contains 2.00 kg, how much would you pay for a case of sucrose containing 12 bottles?

2. An 1859 book, *The Corner Cupboard*, lists prices and common measurements in use at that time. If a *firkin* of beer costs $1.39, how much would you pay for a *tierce* of beer? (1 firkin = exactly 9 gal and 1 tierce = exactly 42 gal)

3. Your lab bench is 36 inches by 120 inches. How many square meters of acid resistant film will you need to cover the bench? (Use 1.0 m = 39 inches.)

Activity 01-2

Unit Analysis

Why?

Unit analysis (aka dimensional analysis) is a procedure that produces the units associated with answers to mathematical calculations. It facilitates problem solving, validates the solutions, and sometimes involves unit conversions. Engineers, health professionals, biologists, and other scientists often rely on unit analysis in their work.

Learning Objective

- Master the use of dimensional analysis in making unit conversions, solving problems, and validating answers.

Success Criterion

- Set up, compute, and validate the solution (units and the magnitude) of computational problems.

Prerequisite

- **Activity 01-1:** *Units of Measurement*

Model: A General Chemistry Problem-Solving Strategy Applied to Dimensional Analysis

In the country of Apocrib, force is measured in units of nocums and area is minum squared. Find the pressure (both the units and the magnitude) when a force of 12 nocums is exerted on a window pane with an area of 24 minum^2. Pressure is defined as force divided by area, P = F/A.

Methodology	Example
Step 1: Identify and record what is known or given.	**Known:** force = 12 nocums area = 24 minum^2
Step 2: Identify and record what is unknown and needs to be found.	**Unknown:** pressure
Step 3: Identify and record the concepts that connect what needs to be found to what is known.	**Connections:** pressure = force / area P = F/A
Step 4: Set up the solution using the connections.	**Setup:** P = F/A P = 12 nocums / 24 minum^2

(continued on the following page)

With contributions by Vicky Minderhout, Seattle University

MODEL: A GENERAL CHEMISTRY PROBLEM-SOLVING STRATEGY APPLIED TO DIMENSIONAL ANALYSIS (CONTINUED)

Methodology	Example
Step 5: Do the mathematics to obtain the result.	**Result:** apply math to both the numbers and the units, obtain 0.50 nocums / minum^2
Step 6: Check or validate your answer.	**Validation:** Units are correct, value is correct (12 / 24 = 0.50)

KEY QUESTIONS

1. What is the purpose of each step in the model problem solving strategy?

what you know
what you don't know

2. In a calculation, how can you obtain the units to associate with the numerical answer?

unit analysis

3. How can you use unit analysis to identify whether you have performed numerical algebraic operations correctly?

everything canceles out and only the unit you wanted is left

4. What meaning is lost when units are omitted? What are the general implications of not using units?

lose your reference point

EXERCISES

1. Density is mass divided by volume. If mass is measured in units of vectas and volume is measured in units of tarts, determine the units of density in this unit system.

   vectas / tarts

2. Chemists have found that PV = nRT for gases where P = pressure in atmospheres, V = volume in liters, n = amount of gas in moles, R = a constant, and T = temperature in Kelvin. Identify the units of the constant R.

   $R = \frac{PV}{nT}$

3. Pressure is defined as force per unit area (P = F/A), and force is defined as mass times acceleration (F = ma). In the SI system, pressure is measured in units of Pascal (Pa). If the following SI units are used: mass = kg, acceleration = m/s^2, and area = m^2, write the dimensional equality statement relating 1 Pa to the SI units of kg, m, and s.

   $Pa = \frac{kg}{ms^2}$

   $P = F/A$

   $P = \frac{ma}{A}$ $P = \frac{mas}{m/s^2}$ $= P = \frac{kg}{m/s^2}$

4. Explain how unit analysis validates your answers to the following.

   a) The size of a diamond is measured in carats. What is the volume of a 1.0 carat diamond if its density is 3.50 g/cm^3? (1 carat = 0.200 g)

   $\frac{1\ carat}{3.50g/cm^3}$ $\frac{2.}{1\ carat}$ | $\frac{1\ carat}{2g}$ $\frac{3.50g}{1\ cm^3}$

   .057 cm^3

   b) In 1973 Secretariat set the current record of 1 minute 59.40 seconds for the Kentucky Derby. The length of this race is 10 furlongs. What was Secretariat's average speed in this race in mi/hr? (1 furlong = 1/8 mile)

   ~~10 furlong~~

   10 furlong $\frac{\frac{1}{8}\ mile}{1\ furlong}$ $\frac{1}{.03317}$ = 37.7 mi/hr

Information

You will find the following information helpful in solving the problem below.

Sodium fluoride, NaF, is 45.0% fluoride by mass.

1 ppm means "one part per million" by mass, e.g., 1 g of fluoride per one million grams of water.

1 gallon = 3.79 L

1 year = 365 days

1 ton = 2000 lb

1 lb = 453.6 g

density of water = 1.0 g/mL

Problem

You are the public health officer in the water treatment facility for a city of 50,000 people. A concentration of 1.0 ppm of fluoride in the drinking water is sufficient for the purpose of helping to prevent tooth decay. The compound normally chosen for fluoridation is the same as is found in some toothpastes, sodium fluoride, NaF. Calculate how many kilograms of sodium fluoride you will need to purchase in order to fluoridate the city's water supply for one year, based upon your estimate that the average daily consumption of water is 150 gallons per person.

$50,000 \text{ people} \; \frac{150 \text{ gallons}}{1 \text{ person}} \; \frac{3.79 \text{ L}}{1 \text{ gallon}} \; \frac{1 \text{ g}}{1\times10^{-3} \text{ mL}} \; \frac{1 \text{ g F}}{1,000,000 \text{ g}} \; \frac{45}{1 \text{ g F}}$

22630 kg

$50,000 \left(\frac{150 \text{ gallons}}{1 \text{ person}}\right) = 7.5\times10^{6} \text{ gal}$

$7.5\times10^{6} \text{ gal} \; \frac{3.79 \text{ L}}{1 \text{ gall}} = 2.8\times10^{7} \text{ L}$

$2.8\times10^{7} \text{ L} \left(\frac{1000 \text{ mL}}{1 \text{ L}}\right) \frac{1 \text{ g}}{1 \text{ mL}} = 2.8\times10^{10} \text{ grams}$

$2.8\times10^{10} \left(\frac{1 \text{ g F}}{1\times10^{6} \text{ g}}\right) = 2.8\times10^{4} \text{ g F}$

$2.8\times10^{4} \text{ g F} \; \frac{100 \text{ g NaF}}{45 \text{ g F}} = 6.2\times10^{4} \text{ g NaF}$

Activity 01-3

Significant Figures in Data

Why?

The number of digits or figures reported for a numerical quantity conveys the quality of the measurement or result to the reader. For example, a loan interest rate of 8.00% or 8.25% cited by a salesperson appears to be more exact than one quoted as 8%. In any business involving numerical values, the quality of the value is vital information. In this course and in others, and in your career, you will be expected to use a meaningful number of digits in reporting your numerical results.

Learning Objectives

- Appreciate the difference between accuracy and precision.
- Understand the relationship between precision and the number of significant figures in a number.

Success Criteria

- Identify the accuracy and precision of a numerical value.
- Report computed values to the correct number of significant figures.

Prerequisites

- **Activity 01-1:** *Units of Measurement*
- **Activity 01-2:** *Unit Analysis*

Information

Accuracy is the degree of conformity to a standard or true value.

Precision is the smallest repeatable digit of a measurement.

Significant figures are the repeatable digits and the first uncertain digit in a measurement or calculation.

Model 1: Accuracy, Precision, and Significant Figures

Item	Values	Significant Figures
Bureau of Standards Time	9:15:13.004	8
Jerry's analog watch	9:15	3
Jennifer's digital watch	9:17:52	5
average mass of gold coin	23.32 g	4
height of an index card	0.0770 m	3

Key Questions

1. In measuring time, which value in the model represents the standard value?

Standard time

2. Is Jerry's or Jennifer's watch more accurate? Explain.

Jerry closer to standard

3. Is Jerry's or Jennifer's watch more precise? Explain.

Jennifers because it measures to a smaller incramite

4. How is precision represented in reporting a measurement?

number of sig fig

5. What are two ways to improve
   a) the precision of a measurement?

   more sig fig

   b) the accuracy of a measurement?

   proper measure technique

Exercises

1. Specify the number of significant figures in a) through f) below. Identify any cases where there may be ambiguity in the number of significant figures.
   a) 101.1 4
   b) 0.0125 3
   c) 100 1
   d) 1.00×10^2 3
   e) 100. (Note: including the decimal point is a convention to show that the number is 100 and not 101 or 102)

   3

   f) 0.005700 4

2. Express the number 500 so it is clear that there are only two significant figures.

5.0×10^2

PROBLEMS

1. The mass of a gold coin was measured three times and each measurement was made to five digits. The mass values were 23.319 g, 23.341 g, and 23.296 g. The average mass was reported as 23.32 g. The actual mass of the coin is 25.5631 g.

   a) Are these measurements precise? Explain your answer.

   yes cause they are close to eachother

   b) Are these measurements accurate? Explain your answer.

   No around 2 grams off

   c) Why is the average mass of the gold coin reported to only four significant figures?

   to round it

MODEL 2: ARITHMETIC OPERATIONS AND SIGNIFICANT FIGURES

Addition

23.26	g
100.1	g
0.03631	g
123.39631	g

report as: **123.4 g**

Subtraction

45.8	g
- 3.26	g
42.54	g

report as: **42.5 g**

Multiplication

30.21 m × 27 m = 815.67 m^2

report as: **820 m^2**

Division

4.203 m / 0.0920 s = 45.6847826087 m/s

report as: **45.7 m/s**

Detailed Consideration of the Multiplication and Division Examples

In the multiplication example, the multiplicand, 27 m, has the smaller number of significant figures, two. The value 27 m is uncertain by 1 unit. It could be 26 m, it could be 28 m. Examine multiplication with these two possibilities.

30.21 m × 26 m = 785.46 m^2

30.21 m × 28 m = 845.88 m^2

The three results differ in the tens place, so the product of the two numbers is reported to only two significant digits, 820 m^2. Notice that the product has the same number of significant figures as the *least* certain multiplicand.

In Exercise 3 below, you will be asked to analyze the division example in a way similar to the example in **Model 2** to show that the quotient should be reported with three significant figures.

Key Questions

6. When you add or subtract numbers as shown in **Model 2**, how can you identify the first uncertain digit in the result?

answer has as many as measurement with fewest sig figs

7. When you multiply or divide numbers, what is the relationship between the number of significant digits in the result and the number of significant figures in the numbers you are multiplying or dividing as illustrated in **Model 2**?

answer has as many sig figs as number with least sigfigs being multiplied

Exercises

3. Conduct a detailed consideration of the division example in **Model 2** just as was done for the multiplication example to show that the result for 4.203 m / 0.0920 s should be reported to only three significant figures.

three is smallest number of sig fig in the problem 0.0920 has 3 sig fig becaust leading are not sig figs

45.7

4. Report the total mass of three people weighing 53 kg, 60.4 kg, and 75.67 kg. Explain the rationale for the number of significant figures in your answer.

189.07 = 190 because 53 only has 2 sig figs

5. Calculate the density (mass/volume) of a coin with a mass of 8.4 g and a volume of 0.942 cm^3. Explain why you should report the result to two significant figures and not to one or three.

8.9 because 8.4 only has 2 sigfigs so answer should only have 2

Activity 02-1

Atoms, Isotopes, and Ions

Why?

Atoms are the fundamental building blocks of all substances. To begin to understand the properties of atoms and how they combine to form molecules, you must be familiar with their composition and structure.

Learning Objectives

- Understand the composition and structure of atoms, isotopes, and ions
- Understand how atomic symbols and names identify the number of particles composing an atom, isotope, or ion

Success Criteria

- Use atomic symbols to represent different isotopes and ions
- Given one or more of the following items, determine the others: name, atomic symbol, atomic number, mass number, neutron number, and electron number
- Calculate the percent of the atomic mass that is located in the nucleus of an atom
- Compare the size of an atom to the size of the atomic nucleus

Prerequisites

- Calculation of percent
- **Activity 01-1:** *Units of Measurement*
- **Activity 01-2:** *Unit Analysis*

Information

Matter, which is anything that has mass and occupies space, is composed of substances and mixtures of substances.

A *substance*, or more explicitly, a *pure substance*, is a variety of matter that has uniform and constant composition. For example, pure water is a substance.

Mixtures are composed of two or more substances. For example, salt water is a mixture, even though it is uniform, because the amount of salt in the water (the composition) can vary.

An *element* is a substance that cannot be decomposed into two or more other substances by chemical or physical means. In nuclear reactions, however, one element can be converted into one or more other elements. Only about 118 different elements are known to exist.

An *atom* is the smallest part of an element that can exist either alone or in combination with other atoms.

Isotopes are atoms that have the same number of protons but different numbers of neutrons.

An *ion* is an atom or molecule with a positive or negative charge.

A *cation* is an ion with a positive charge.

An *anion* is an ion with a negative charge.

MODEL: SODIUM

The diagrams below show representations of sodium. Note that the diameter of an atom is about 10,000 times larger than the diameter of the atomic nucleus.

Table 1 Subatomic Particles

Particle	Mass (amu)*	Charge
Proton	1.0073	+1
Neutron	1.0087	0
Electron	0.0005	–1

* *Atomic mass unit* (amu) is a unit of mass equal to 1.66054×10^{-27} kg.

Atomic Symbol Notation

Figure 1

$^{A}_{Z}X^{+}$

Mass Number	Charge
Atomic Symbol	
Atomic Number	

Sodium-23

$^{23}_{11}\text{Na}$

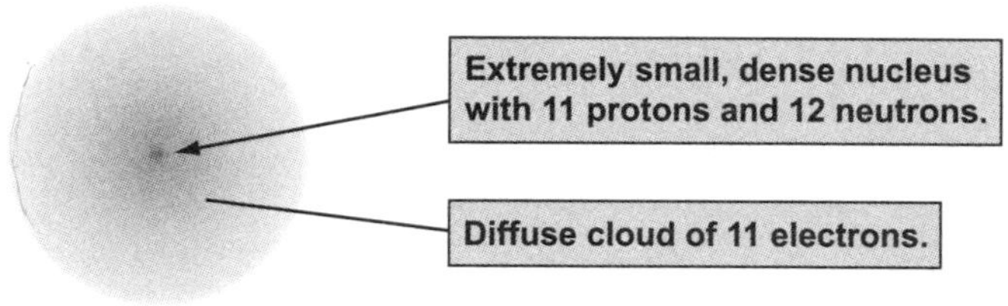

Sodium-24

$^{24}_{11}\text{Na}$

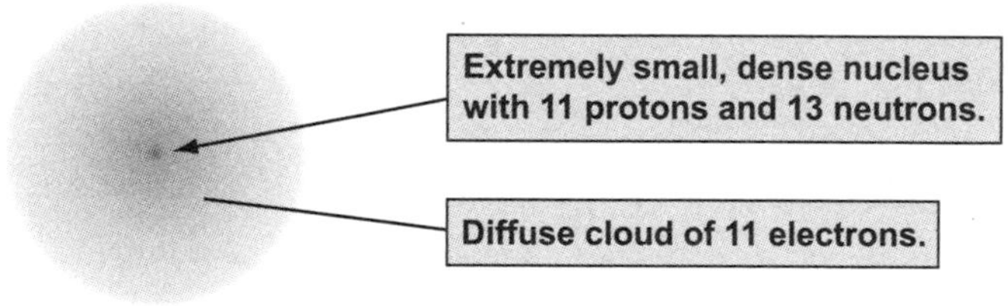

Sodium-24 cation

$^{24}_{11}\text{Na}^{+}$

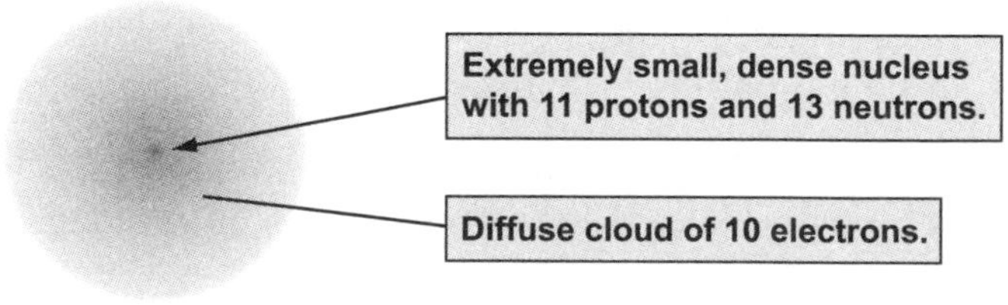

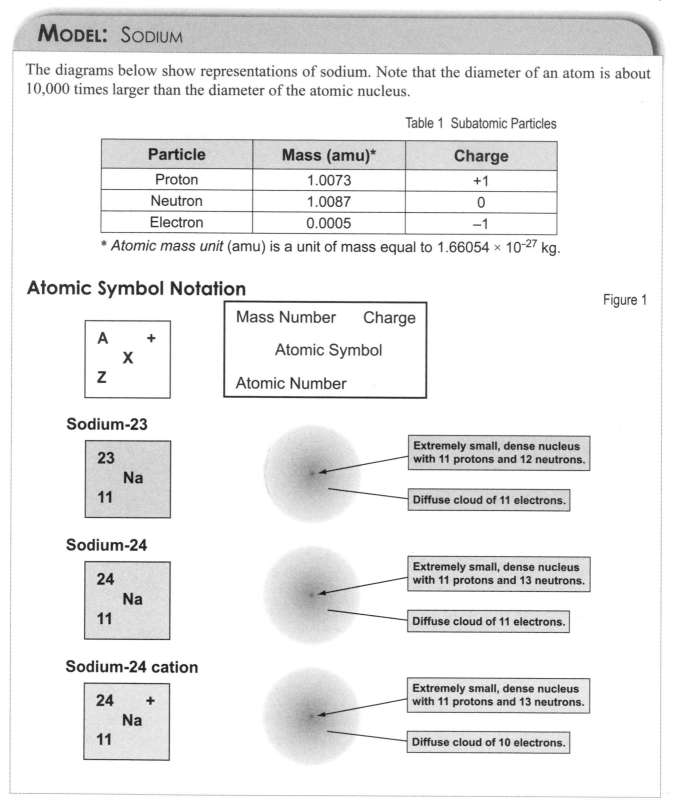

KEY QUESTIONS

1. What are the three particles that comprise a sodium atom?

   proton Neutron Electron

2. Which particles contribute most of the mass to the atom, and where are these particles located?

   Proton and Neutron nucleus

3. Which particles contribute most to the volume or size of the atom, and where are these particles located? electrons around the nucleus

4. What information is provided by the atomic number, Z, which is the subscript in the atomic symbol?

number of electrons

5. What information is provided by the mass number, A, which is the superscript in the atomic symbol? number of protons and neutrons added together

6. What notation is used in the atomic symbol to indicate the charge of an atom or ion?

+ or − sign

7. Given the definition of mass number and the information in Table 1 regarding the masses of protons, neutrons, and electrons, why is the mass number approximately, but not exactly equal to, the mass of an atom in amu? it is rounded and not an exact number

8. How is the charge of the atom or ion determined from the number of protons, neutrons, and electrons present? if there are more electrons it is negative
if there are more protons it is positive

9. What do all atoms and ions of sodium have in common?

either 11 protons

10. In general, what feature of an atom identifies it as a particular element?

electrons

11. In general, how do isotopes of the same element differ?

different number of neutrons

12. How many isotopes of any particular element could there be? What might prevent all of these isotopes from occurring naturally? many but it may not be natural because it would change the element

EXERCISES

1. Insert the missing information in the following table. The first row is completed for you to provide an example.

Table 2

Name	Symbol	Z	A	Number of Neutrons	Number of Electrons
boron-10	$^{10}_{5}B^{+}$	5	10	5	4
calcium-40	$^{40}_{20}Ca$	20	40	20	18
oxygen-16	$^{16}_{8}O$	8	16	8	8
uranium-238	$^{238}_{92}U$	92	238	146	92
Fluorine-19	$^{19}_{9}F$	9	19	10	9
chlorine-35	$^{35}_{17}Cl$	17	35	18	18
Potassium-39	$^{39}_{19}K$	19	39	20	19

2. Show how to calculate the mass of a proton, neutron, and electron in kilograms using the data in Table 1 and the equality statement: **1 amu = 1.66054 × 10^-27 kg**

B

proton 1.0073amu $\frac{1.66054 \times 10^{-27}}{1 amu}$

neutron 1.0087amu $\frac{1.66054 \times 10^{-27}}{1 amu}$

electron .0005amu $\frac{1.66054 \times 10^{-27}}{1 amu}$

PROBLEMS

1. The mass of a carbon-12 atom is 12 amu. What percent of the mass is located in the nucleus? Why is the value you calculated so close to 100%?

12 amu

% mass in nucleus $\left(\frac{\text{mass of nucleus}}{\text{mass of atom}}\right) \times 100$

$\frac{12 \text{ amu} - 6(.0005 \text{ amu})}{12 \text{ amu}}$

2. The radius of a Cl nucleus is 4.0 fm, and the radius of a Cl atom is 100 pm. If the nucleus of the Cl atom were the size of a dime, which is 17 mm in diameter, determine whether the atom would be approximately a) the size of a quarter, b) the size of a car, c) the size of a football stadium, or d) the size of the earth. Explain how you made your decision.

nucleus 4.0 fm = 4.0×10^{-15} m

atom 100 pm = 100×10^{-12} m

$\frac{\text{atom}}{\text{nucleus}} = \frac{x}{\text{dime}}$ $\frac{100 \times 10^{-12}}{4.0 \times 10^{-15}} = \frac{x}{17\text{mm}}$ x = 425 m

Activity 02-2

Mass Spectrometry and Masses of Atoms

Why?

The mass of an atom in amu is approximately equal to its mass number, but a powerful instrument called a mass spectrometer can be used to determine the exact masses of atoms, reveal the presence of isotopes, and measure the naturally occurring proportion of each isotope. With this information, the average mass of all the isotopes of an element can be determined, and then you can count the number of atoms or molecules present in a sample by weighing it. Knowing how many atoms or molecules are in a sample is essential to understanding chemical reactions and using chemical compounds and reactions in research, industrial technology, and medicine.

Learning Objective

- Understand the basic idea of mass spectrometry, and how the mass of an atom can be determined

Success Criteria

- Successfully interpret a mass spectrum
- Successfully construct a mass spectrum from appropriate data
- Convert between mass in amu and mass in grams
- Determine the number of atoms in a given mass of an element

Prerequisites

- **Activity 01-1:** *Units of Measurement*
- **Activity 01-2:** *Unit Analysis*
- **Activity 01-3:** *Significant Figures in Data*
- **Activity 02-1:** *Atoms, Isotopes and Ions*

Information

In a mass spectrometer, an electrical discharge knocks electrons off the atoms or molecules giving them a net positive charge. These ions then are accelerated in an electric field, and some property related to their mass is measured. In some mass spectrometers, the property is the trajectory of the ions in a magnetic field. In other mass spectrometers, the property is the time taken by the ions to move the distance from the point of ionization to a detector. An example of a mass spectrum is shown in **Model 1**.

In a mass spectrum the number of ion counts over some period of time is plotted on the y-axis and the atomic mass is plotted on the x-axis. The number of ion counts for different isotopes is proportional to the abundance of each isotope in the sample. Since atomic masses are so small, a special *mass* unit is used. This unit is called the *atomic mass unit* and is abbreviated *amu*. The *atomic mass unit* is defined so the mass of the carbon-12 isotope is exactly 12 amu. The masses of all other atoms then are measured relative to the mass of carbon-12. The factor for converting amu to grams is 1.66054×10^{-24} g/amu.

MODEL: DATA ON THE MASSES OF ATOMS

Mass Spectrum of Boron

Figure 1

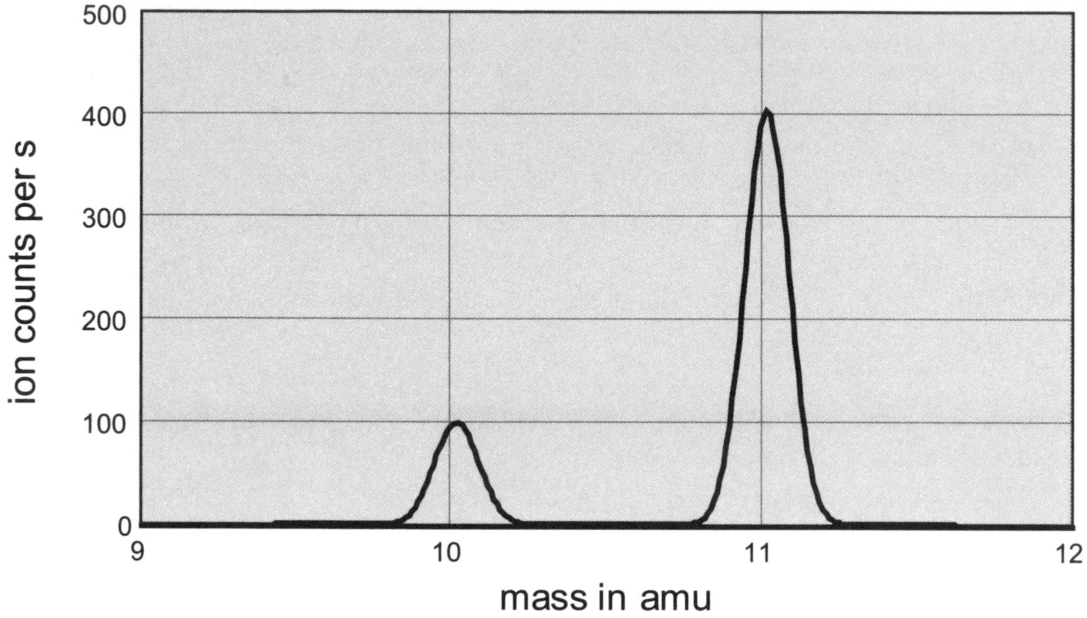

Isotopic Data

Table 1

Atomic Isotope	Symbol	Mass (amu)*	Natural Abundance on Earth (%)
Hydrogen	${}^{1}_{1}H$	1.0078	99.985
Deuterium	${}^{2}_{1}H$	2.0140	0.015
Helium	${}^{4}_{2}He$	4.00260	100.0
Boron-10	${}^{10}_{5}B$	10.0129	19.78
Boron-11	${}^{11}_{5}B$	11.0093	80.22
Carbon-12	${}^{12}_{6}C$	exactly 12 by definition	98.89
Carbon-13	${}^{13}_{6}C$	13.0034	1.11
Chlorine-35	${}^{35}_{17}Cl$	34.9689	75.53
Chlorine-37	${}^{37}_{17}Cl$	36.9481	24.47

* 1 Atomic mass unit (amu) is a unit of mass equal to 1.66054×10^{-24} g.

Key Questions

1. What mass unit is used to report the mass of atoms in the model?

   amu

2. According to the *Information* section, how is this mass unit defined?

   carbon 12 isotope is exactly 12 amu

3. Why are there two peaks in the mass spectrum of boron?

   because its showing 2 isotopes

4. Why is one peak in the mass spectrum for boron higher than the other?

   there ~~are~~ is more Boron 11

5. Are the positions of the peaks on the x-axis and the ratio of peak heights in the boron mass spectrum consistent with the data in *Table 1: Isotopic Data*? Explain.

   yes there is a greater abundance of Boron 11

6. The mass number and atomic mass both have a value of 12 for carbon-12, do isotopes of all other elements also have the same value for the mass number and atomic mass? Explain.

   no

7. Why is it more convenient to use atomic mass units (amu) rather than grams or kilograms (g or kg) in reporting masses of individual atoms?

   because the atomic mass would be very small if it was measured in g or kg

8. If an element has several isotopes, what single value for the mass could be used to characterize atoms of that element? Justify your answer.

   atomic mass

Exercises

1. Calculate the ratio of boron-11 to boron-10 found on Earth using the information given in Table 1.

   $\frac{80.22}{19.78}$

2. Describe how you can determine the ratio of boron-11 to boron-10 found on Earth from information provided by the mass spectrum of boron.

3. From the information in Table 1, determine the number of boron-11 atoms that you would expect to find in a natural sample of 10,000 boron atoms.

4. Draw a sketch to show what you expect the mass spectrum of a natural sample of atomic chlorine to look like.

5. If you could pick one carbon atom from a natural sample, what would its mass in grams most likely be? Explain.

Figure 2

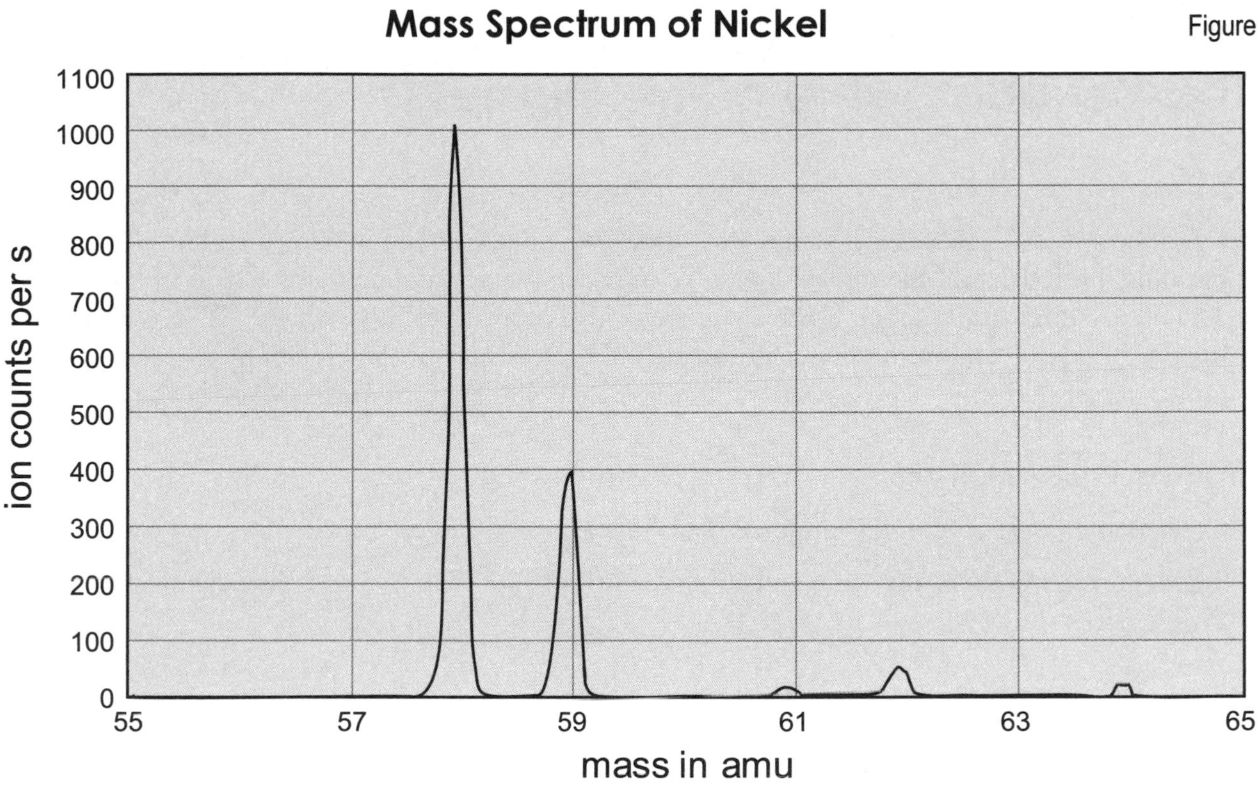

6. From the mass spectrum for nickel shown in Figure 2, determine
   a) the number of nickel isotopes present in the sample.

   b) the mass numbers of the nickel isotopes.

   c) the number of neutrons in each isotope.

   d) the relative abundance of each isotope.

Activity 02-3

The Periodic Table of the Elements

Why?

Substances that contain only atoms with the same number of protons are called *elements*. The Periodic Table lists all the known elements in order of their atomic number and in columns that depend on similarities in their chemical and physical properties. The Periodic Table is a useful tool for both students and professionals to identify the properties of the elements and understand the properties of molecules.

Learning Objectives

- Become familiar with the organization of the Periodic Table
- Appreciate both the diversity and commonalities in the chemical and physical properties of the elements

Success Criteria

- Identify groups and periods in the Periodic Table
- Use the Periodic Table to provide information about the elements

Prerequisites

- **Activity 02-1:** *Atoms, Isotopes and Ions*
- **Activity 02-2:** *Mass Spectrometry and Masses of Atoms*

Information

Dmitri Mendeleev (1834 – 1907), a Russian scientist, constructed the first Periodic Table by listing the elements in horizontal rows in order of increasing atomic mass. He started new rows whenever necessary to place elements with similar properties in the same vertical column. Mendeleev found that the correlations in properties between some elements in the columns were not perfect. These observations led him to predict the existence of undiscovered elements and to wonder how the table might be better organized. Later H.G.J. Moseley used x-ray spectra to refine the ordering and show that atomic numbers rather than atomic masses should be used to order the elements.

In the Periodic Table, elements with similar properties occur in vertical columns called *groups*. Two numbering conventions are used to label the groups. The older convention numbers the groups using Roman numerals I through VIII followed by a letter A or B; the other convention numbers each column 1 through 18. The A groups are known as the *main group elements*. The B groups are called the *transition elements*. The group numbers IA through VIIIA in the older convention tells you how many valence electrons an element has. The valence electrons are the outer electrons that are most important in determining the chemical bonding and other properties of the element.

The horizontal rows of the table are called *periods*, and are numbered 1 through 7 starting with the row that only contains H and He.

MODEL: THE PERIODIC TABLE OF THE ELEMENTS

The Periodic Table of the Elements

Key: 1.008 — Mean atomic mass (amu) / Molar Mass (g/mol); H — Symbol; Hydrogen — Name; 1 — Atomic number

Period	1 IA	2 IIA	3 IIIB	4 IVB	5 VB	6 VIB	7 VIIB	8 VIIIB	9 VIIIB	10 VIIIB	11 IB	12 IIB	13 IIIA	14 IVA	15 VA	16 VIA	17 VIIA	18 VIIIA
1	1.008 H Hydrogen 1																	4.00 He Helium 2
2	6.94 Li Lithium 3	9.01 Be Beryllium 4											10.81 B Boron 5	12.01 C Carbon 6	14.01 N Nitrogen 7	16.00 O Oxygen 8	19.00 F Fluorine 9	20.18 Ne Neon 10
3	22.99 Na Sodium 11	24.31 Mg Magnesium 12											26.98 Al Aluminum 13	28.09 Si Silicon 14	30.97 P Phosphorus 15	32.07 S Sulfur 16	35.45 Cl Chlorine 17	39.95 Ar Argon 18
4	39.10 K Potassium 19	40.08 Ca Calcium 20	44.96 Sc Scandium 21	47.88 Ti Titanium 22	50.94 V Vanadium 23	52.00 Cr Chromium 24	54.94 Mn Manganese 25	55.85 Fe Iron 26	58.93 Co Cobalt 27	58.69 Ni Nickel 28	63.55 Cu Copper 29	65.39 Zn Zinc 30	69.72 Ga Gallium 31	72.61 Ge Germanium 32	74.92 As Arsenic 33	78.96 Se Selenium 34	79.90 Br Bromine 35	83.80 Kr Krypton 36
5	85.47 Rb Rubidium 37	87.62 Sr Strontium 38	88.91 Y Yttrium 39	91.22 Zr Zirconium 40	92.91 Nb Niobium 41	95.94 Mo Molybdenum 42	(97.9) Tc Technetium 43	101.07 Ru Ruthenium 44	102.91 Rh Rhodium 45	106.42 Pd Palladium 46	107.87 Ag Silver 47	112.41 Cd Cadmium 48	114.82 In Indium 49	118.71 Sn Tin 50	121.76 Sb Antimony 51	127.60 Te Tellurium 52	126.90 I Iodine 53	131.29 Xe Xenon 54
6	132.91 Cs Caesium 55	137.33 Ba Barium 56	138.91 La Lathanum 57	178.49 Hf Hafnium 72	180.95 Ta Tantalum 73	183.85 W Tungsten 74	186.21 Re Rhenium 75	190.2 Os Osmium 76	192.22 Ir Iridium 77	195.08 Pt Platinum 78	196.97 Au Gold 79	200.59 Hg Mercury 80	204.38 Tl Thallium 81	207.2 Pb Lead 82	208.98 Bi Bismuth 83	(209) Po Polonium 84	(210) At Astatine 85	(222) Rn Radon 86
7	(223.02) Fr Francium 87	(226.03) Ra Radium 88	(227.03) Ac Actinium 89	(261) Rf Rutherfordium 104	(262) Db Dubnium 105	(263) Sg Seaborgium 106	(262) Bh Bohrium 107	(265) Hs Hassium 108	(266) Mt Meitnerium 109	(271) Ds Darmstadtium 110	(280) Rg Roentgenium 111	(285) Not Yet Officially Named 112						
	Alkali Metals	Alkaline Earth Metals	Transition Metals														Halogens	Noble Gases

Lathanides	140.12 Ce Cerium 58	140.91 Pr Praseodymium 59	144.24 Nd Neodymium 60	(145) Pm Promethium 61	150.36 Sm Samarium 62	152.97 Eu Europium 63	157.25 Gd Gadolinium 64	158.93 Tb Terbium 65	162.50 Dy Dysprosium 66	164.93 Ho Holmium 67	167.26 Er Erbium 68	168.93 Tm Thulium 69	173.04 Yb Ytterbium 70	174.97 Lu Lutetium 71
Actinides	232.04 Th Thorium 90	(231.04) Pa Protactinium 91	238.03 U Uranium 92	(237.05) Np Neptunium 93	(240) Pu Plutonium 94	(243.06) Am Americium 95	(247) Cm Curium 96	(248) Bk Berkelium 97	(251) Cf Californium 98	(252.08) Es Einsteinium 99	(257.10) Fm Fermium 100	(257) Md Mendelevium 101	(259.10) No Nobelium 102	(262.11) Lr Lawrencium 103

Legend: Non-Metals (dark shading); Metals (light shading); Semi-Metals (white)

Information

There are three categories of elements in the Periodic Table: metals, nonmetals, and metalloids. The metals are located in the left and center. They are good conductors of heat and electricity. The nonmetals are in the upper right-hand corner. They are poor conductors of heat and electricity. The metals and nonmetals are separated by the metalloids, which are six elements on a diagonal line. These elements are B, Si, Ge, As, Sb, and Te. The metalloids are also called semimetals or semiconductors because their conductivity is between that of metals and nonmetals. Metals readily lose electrons to form positive ions, called cations, and nonmetals readily gain electrons to form negative ions, called anions.

Key Questions

1. What information about an element is provided in each box for that element in the Periodic Table in the model?

   Atomic mass

   Atomic number name symbol

2. What determines the sequence of the elements from the first to the last?

   number of protons

3. What determines where one row stops and another begins?

   from noble gass to metal
   Alkali

4. Where are the metals, nonmetals, and metalloids located?

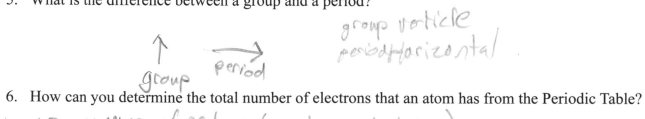

5. What is the difference between a group and a period?

   group period

   group verticle
   period horizontal

6. How can you determine the total number of electrons that an atom has from the Periodic Table?

   Atomic # = number of protons (number at botom)
   6.41 Li
   3

7. How can you determine the number of valence electrons that atoms in groups 1, 2 and 13 through 18 have?

   The Roman numeralk

8. What are the five other elements like helium that are gases and are not very reactive?

Exercises

1. Write the name, symbol, atomic number, and average mass for the Group 2 metal in Period 3.

2. Write the name, symbol, average mass, and number of protons for the Group 16 nonmetal in Period 2.

3. Write the name, symbol, average mass, and number of electrons for the Group 15 metalloid in Period 4

4. Write the name and symbol of the element that has 48 electrons.

5. Name two elements that have properties similar to sodium, Na. How many valence electrons do each of these three elements have?

6. Name two elements that have properties similar to bromine, Br. How many valence electrons do each of these three elements have?

7. Using atomic symbols, list the elements in Period 2 in order of increasing number of electrons.

8. Using atomic symbols, list the elements in Group 14 in order of increasing number of protons.

9. Using atomic symbols, list the elements in Group 12 in order of increasing atomic mass.

ACTIVITY 03-1

Molecular Representations

WHY?

Molecules are formed from collections of atoms and are represented symbolically in several different ways. These representations are part of the language of chemistry and convey essential information about the molecule. Being able to construct these representations and interpret them is essential to learning and understanding chemistry.

LEARNING OBJECTIVE

- Represent molecules by their molecular and structural formulas, and by line drawings

SUCCESS CRITERION

- Interconvert among molecular formulas, structural formulas, and line drawings

PREREQUISITE

- **Activity 02-1:** *Atoms, Isotopes, and Ions*

INFORMATION

Atoms combine to form molecules because molecules are more stable than atoms. By more stable, we mean that they have a lower energy. The forces that hold the atoms together are called *chemical bonds*. These forces are due to the attraction between the negatively charged electrons and the positively charged atomic nuclei. In the simplest picture, a chemical bond consists of two electrons located between two atomic nuclei.

Molecular formulas (also called *chemical formulas*) designate the composition of a molecule by giving the number of atoms of each element in the molecule. For example, the chemical formula H_2O means that two atoms of hydrogen combine with one atom of oxygen to produce water. In most cases a molecular formula alone does not uniquely identify a molecule. One must also identify the molecular structure, the three-dimensional manner in which the atoms are bonded together.

Structural formulas show the individual bonds between the atoms within a molecule represented as lines. A *line drawing* is a special type of structural formula in which the carbon atoms, and hydrogen atoms bound to carbon atoms, are not shown. Lines are used to represent the carbon-carbon bonds with the proper bond angles. *Line drawings* are much superior to structural formulas because they provide some information on molecular geometry and are easier to draw!

Contributed by Joseph Lauher, Stony Brook University
Modified by David Hanson, Stony Brook University

MODEL: EXAMPLES OF STRUCTURAL FORMULAS AND LINE DRAWINGS

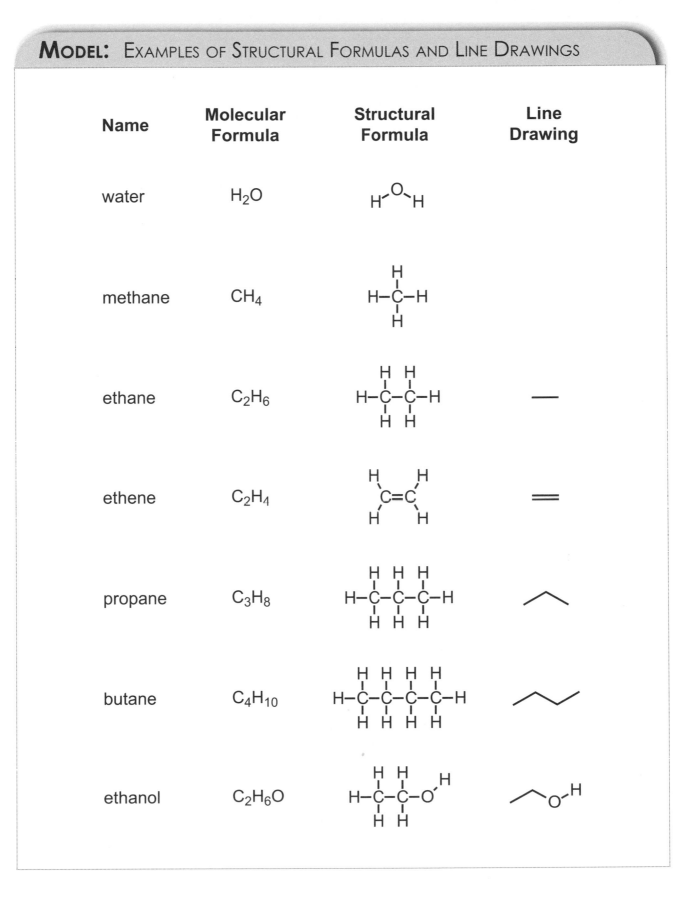

Name	Molecular Formula	Structural Formula	Line Drawing
water	H_2O		
methane	CH_4	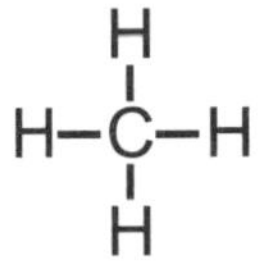	
ethane	C_2H_6	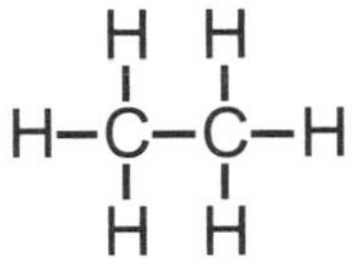	
ethene	C_2H_4	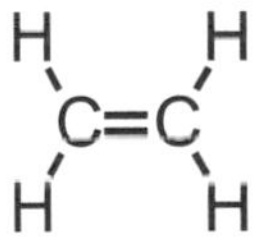	
propane	C_3H_8	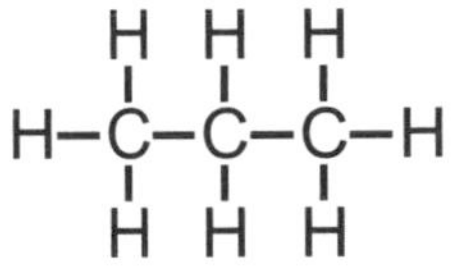	
butane	C_4H_{10}	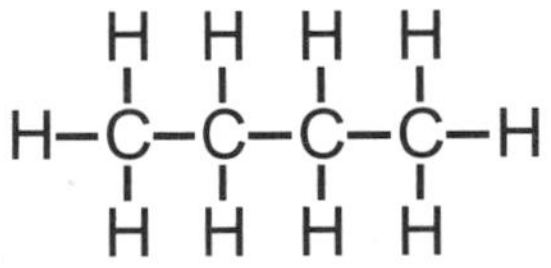	
ethanol	C_2H_6O	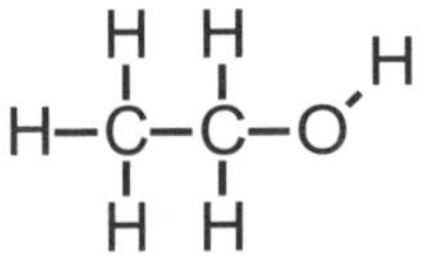	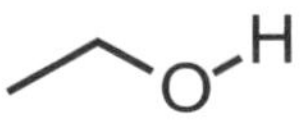

Key Questions

1. What is the molecular formula for propane?

2. What is the structural formula for ethene?

3. What does a line drawing represent?

4. What do the lines in the structural formulas and line drawings represent?

5. What element is present at the end of a line in a line drawing even though no atomic symbol is given?

6. How many bonds does each carbon atom form as shown by the structural formulas?

7. How can one determine the number of hydrogen atoms bound to a given carbon atom from a line drawing?

Example

Consider the line structure of the molecule 1-butene: 

Starting on the left, the first carbon has two bonds so it must have two hydrogen atoms. The second carbon has three bonds so it has one hydrogen atom. The third carbon has two bonds and thus two hydrogen atoms and the fourth and final carbon has only one bond and must have three hydrogen atoms. Adding up we find that the molecular formula is C_4H_8 and the structural formula is:

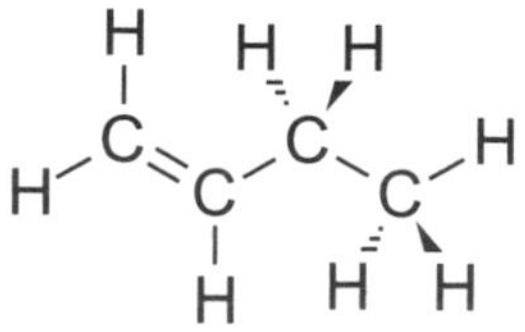

EXERCISE 1

In the spaces provided below, write the molecular and structural formulas for each of the following line drawings. Note that the structural formula explicitly includes all carbon and hydrogen atoms.

benzene

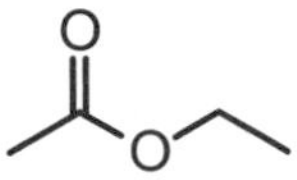

ethylacetate

acetone

3-hexyne

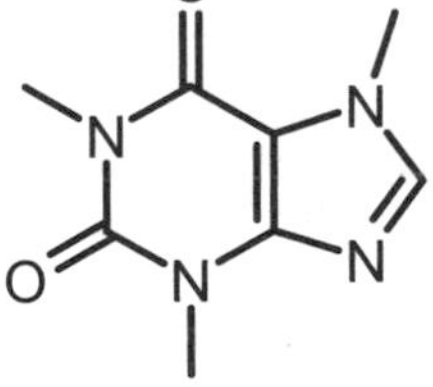

caffeine

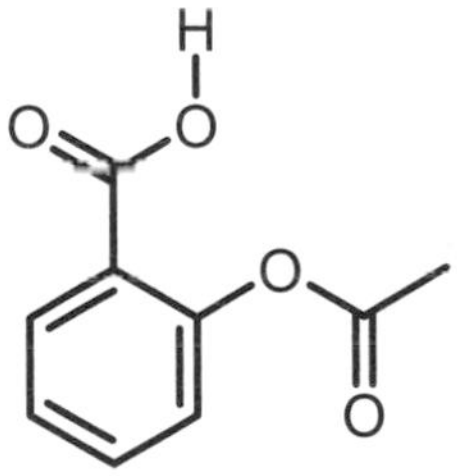

aspirin

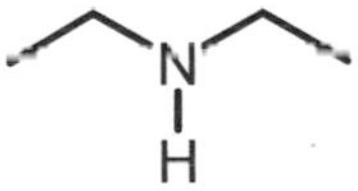

diethylamine

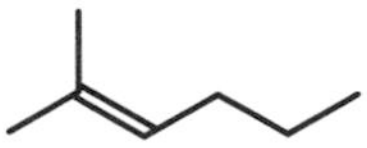

2-methyl-2-hexene

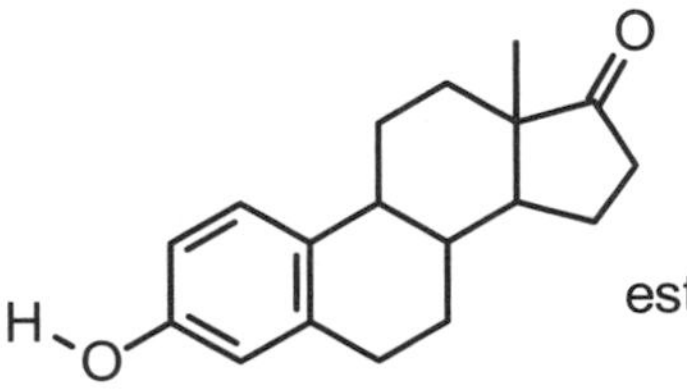

estrone

Activity 03-2

Nomenclature: Naming Compounds

Why?

The term *nomenclature* refers to a system of principles, procedures, and terms for naming things. Names are an intrinsic part of the way we communicate information. So in order to talk about chemical compounds, which are substances formed from two or more different elements, in a meaningful way, they need to have names that that tell us something about their composition or molecular formula.

Learning Objective

- Understand and apply the nomenclature for binary ionic compounds, binary covalent compounds, strong acids, and polyatomic ions

Success Criteria

- Correctly write the molecular formula when given the name of a compound
- Correctly name a compound when given its molecular formula

Prerequisites

- **Activity 02-3:** *Periodic Table of the Elements*
- **Activity 03-1:** *Molecular Representations*

Information

Electronegativity is a measure of the ability of an atom to attract electrons in a molecule. A chart of electronegativity values is provided on the back cover of this book. Electrons in a bond are pulled toward the atom that has the larger electronegativity. If the electronegativities of two elements are not too different, which generally means a difference less than or equal to 1.9, then electrons are shared between the two atoms, and the compound is called covalent. If the difference in electronegativities is greater than 1.9, then the more electronegative atom attracts one or more electrons from the other atom, and the compound is called ionic.

Model 1: Names of Binary Covalent Compounds

Molecular Formula	Name
HCl	Hydrogen chloride
H_2S	Hydrogen sulfide
$TiCl_4$	Titanium tetrachloride
CO	Carbon monoxide
CO_2	Carbon dioxide

Molecular Formula	Name
PBr_3	Phosphorous tribromide
CCl_4	Carbon tetrachloride
N_2O_5	Dinitrogen pentoxide
NO_2	Nitrogen dioxide
SF_6	Sulfur hexafluoride

Key Questions

1. What does the term *binary covalent* mean?

doesn't contain metal only contains two elements

2. In both the molecular formula and the name, which element is given first, the more electronegative or the less electronegative?

less electronegative

3. What ending is applied to the root of the more electronegative element in the name?

"ide"

4. What prefixes are used to indicate the number of atoms when there is more than one possibility?

mono, di, tri,

Model 2: Names of Binary Ionic Compounds

Molecular Formula	Name
NaCl	Sodium chloride
KBr	Potassium bromide
$MgCl_2$	Magnesium chloride
$CaCl_2$	Calcium chloride
CaO	Calcium oxide

Molecular Formula	Name
$CoCl_2$	Cobalt(II) chloride
$CoCl_3$	Cobalt(III) chloride
FeO	Iron(II) oxide
Fe_2O_3	Iron(III) oxide
Al_2O_3	Aluminum oxide

Key Questions

5. In the molecular formula and the name, which element is given first, the metal or the nonmetal?

6. What ending is applied to the root of the nonmetal in an ionic compound?

7. When a metal ion can form more than one kind of cation, how is the charge on the cation indicated?

roman numerals

8. Since a Roman numeral is not included in the names of the alkali metal and alkaline earth metal compounds, how do you determine the charge on the cation and the number of anions associated with it?

look at subscript of anion

9. What are some similarities and differences in the names of binary covalent and binary ionic compounds?

same { both end in ide; Cation comes first; name the first one }

differences { Roman numerals; prefixes are only in binary covalent }

EXERCISES

1. Use your textbook or other reference resource to make a list of binary covalent compounds that require the use of number prefixes in the name (mono, di, tri, tetra, penta, and hexa).

2. Use your textbook or other reference resource to make a list of metals that require the use of Roman numerals to specifiy the cation charge. Be sure to include the charges that those metal cations commonly have.

3. For the following compounds, identify the compound as **covalent** or **ionic** and complete the table by filling in the missing entries.

Type (ionic or covalent)	Molecular Formula	Name
covalent	CS_2	Carbon disulfide
covalent	N_2O_4	dinitrogen tetroxide
Ionic	Na_2S	Sodium sulfide
Ionic	Cu_2O	Copper (I) oxide
Ionic	$CuCl_2$	Copper (II) chloride

Model 3: Names of Acids

Acids and bases are extremely important chemical compounds. You will learn more about them later. For now you need to recognize the names of some common acids and the information provided by the name.

Molecular Formula	Name
HCl	Hydrochloric acid
H_2SO_4	Sulfuric acid
H_2SO_3	Sulfurous acid

Molecular Formula	Name
HNO_3	Nitric acid
HNO_2	Nitrous acid
H_2CO_3	Carbonic acid

Key Questions

10. When two similar acids exist with different numbers of oxygen atoms, how do the names distinguish between these?

11. What are the names of the acids in **Model 3** that have one hydrogen atom?

12. What are the names of the acids in **Model 3** that have two hydrogen atoms?

Model 4: Names of Polyatomic Ions

Molecular Formula	Name
SO_4^{2-}	Sulfate
SO_3^{2-}	Sulfite
NO_3^-	Nitrate
NO_2^-	Nitrite

Molecular Formula	Name
CO_3^{2-}	Carbonate
NH_4^+	Ammonium
H_3O^+	Hydronium
OH^-	Hydroxide

Key Questions

13. When two similar polyatomic ions exist with different numbers of oxygen atoms, how do the names distinguish between these?

14. How does the name identify a polyatomic ion as having a positive charge?

Exercise

4. Some of the acids listed in **Model 3** can be formed from a polyatomic ion with a similar name. Write the names and the molecular formulas of each polyatomic ion and its corresponding acid below. Identify how the acid is formed from the polyatomic ion in each case.

Activity 03-3

Mole and Molar Mass

Why?

To keep track of the huge numbers of atoms and molecules in samples that are large enough to see, chemists have established a unit of counting called the *mole* (abbreviated mol) and a unit of measure called the *molar mass*, which has units of g/mol. By using the idea of a *mole* and *molar mass*, you will be able to count out specific numbers of atoms or molecules simply by weighing them. This capability is necessary in understanding chemical reaction equations, conducting research in chemistry and biology, and applying chemistry in technology and the health sciences.

Learning Objectives

- Understand the relationship between the mole and Avogadro's number
- Understand the meaning of the molar mass of a substance
- Recognize that the molar mass is an average of all the isotopic masses of an element

Success Criteria

- Quickly convert between the number of atoms, moles, and the mass of a sample by using Avogadro's number and the molar mass appropriately
- Calculate the molar mass from isotopic abundances and isotopic masses

Prerequisites

- **Activity 01-2:** *Unit Analysis*
- **Activity 01-3:** *Significant Figures in Data*
- **Activity 02-1:** *Atoms, Isotopes, and Ions*

Model 1: A Mole is a Counting Unit

1 pair of objects = 2 objects

1 dozen objects = 12 objects

1 gross of objects = 144 objects

1 mole of objects = 6.02214×10^{23} objects

KEY QUESTIONS

1. How many pencils are there in a dozen pencils? 12

2. How many pencils are there in a gross of pencils? 144

3. How many pencils are there in a mole of pencils? 6.022×10^{23}

4. How many atoms are there in a dozen atoms? 12

5. How many atoms are there in a gross of atoms? 144

6. How many atoms are there in a mole of atoms? 6.022×10^{23}

7. In what way are the meanings of the terms *pair*, *dozen*, *gross*, and *mole* similar?

 all words for a set number of something

 In what way are the meanings different?

 1 mole

INFORMATION

The number of objects in a mole (6.02214×10^{23}) is so important in chemistry that is given a name. It is called *Avogadro's number*, which has units of objects /mol.

Avogadro's number is determined by the number of carbon atoms in exactly 12 g of pure carbon-12.

The *molar mass* is the mass of a mole of objects. It has units of g/mol.

1 amu = 1.66054×10^{-24} g

EXERCISES

1. A single carbon-12 atom has a mass of 12 amu by definition of the atomic mass unit. Convert 12 amu to grams, and then calculate the mass in grams of a mole of carbon-12 atoms.

 $1 mol \left(\frac{6.022 \times 10^{23}}{1 mol} \right) 12 amu \quad \frac{1g}{6.022 \times 10^{23} amu}$

 $12 amu \frac{1.66054 \times 10^{-24} g}{1 amu} = 1.9926 \times 10^{-23} g$

 $1 mole \left(\frac{6.022 \times 10^{23} atom}{1 mole} \right) \left(\frac{12.01 amu}{atom} \right) \left(\frac{1.66 \times 10^{-24} g}{amu} \right) 12.0 g$

2. A single oxygen-16 atom has a mass of 15.9949 amu. Convert this mass to grams, and then calculate the mass in grams of a mole of oxygen-16 atoms.

 $15.9944 amu \left(\frac{1.66054 \times 10^{-24} g}{1 amu} \right) \quad 2.656 \times 10^{-25} g$

 $1 mole \left(\frac{6.0224 \times 10^{23} atoms}{1 mole} \right) \left(\frac{16 amu}{atoms} \right) \left(\frac{1.66054 \times 10^{-24} g}{amu} \right) = 16$

3. Based on your results for Exercises 1 and 2, identify the relationship between the numerical values of the mass of an atom in amu and the molar mass in g/mol.

they are the same

MODEL 2: BEADS IN A JAR—AVERAGE MASS OF A MIXTURE OF OBJECTS

Natural samples of most elements are mixtures of different isotopes. The mass of Avogadro's number of atoms in such a sample is not the molar mass of a single isotope but is rather an abundance weighted average of the masses of all the isotopes for that element. Exploration of **Model 2** will guide you in determining these average molar masses.

Table 1

Bead Color	Mass of a Single Bead	Number in the Jar	Percent Abundance
red	2.0 g	50	50%
blue	2.5 g	30	30%
yellow	3.0 g	20	20%

KEY QUESTIONS

8. Which color bead has the largest mass? yellow
9. Which color bead is present in the largest number? red
10. Is the average mass of a bead in the jar equal to 2.5 g, which is [(2.0 + 2.5 + 3.0) ÷ 3]? Explain.

    No because each bead has a different abundance

11. Is the average mass of a bead in the jar greater than or less than 2.5 g? Explain, without doing a calculation; just examine the information in Table 1.
12. How can you calculate the average mass of a bead in the jar using the percent abundance given in Table 1? Provide an explanation, then do the calculation.
13. How can you calculate the molar mass of the beads using the average mass that you calculated in the previous question? Provide an explanation, then do the calculation.

Exercises

4. Using your calculation of the molar mass of beads as a guide, show how to determine the molar mass of boron from the data given in Table 2 below. Remember, molar mass has units of g/mol and 1 amu = 1.66054×10^{-24} g.

Table 2

Isotope	Atomic Mass (amu)	Percent Abundance
boron-10	10.0129	19.78%
boron-11	11.0093	80.22%

5. Compare the number you calculated for the molar mass of boron in Exercise 4 with the number given below the symbol for boron in the Periodic Table. From this comparison, identify the information that is provided by the numbers just below the atomic symbols in the Periodic Table.

6. Calculate the number of atoms in exactly 2 moles of helium.

7. Calculate the number of moles corresponding to 2.007×10^{23} atoms of helium.

8. Calculate the mass in grams of 2.5 moles of argon. The molar mass of argon is 39.95 g/mol.

9. Calculate the number of moles in 75 g of iron. The molar mass of iron is 55.85 g/mol.

10. Calculate the number of atoms in 0.25 moles of uranium.

11. Calculate the mass of 12.04×10^{23} atoms of uranium. The molar mass of uranium is 238.0 g/mol.

Got It!

1. Identify the statement below that is correct and explain why it is correct:
   a) The molar mass of an element divided by Avogadro's number gives the average mass of an atom of that element in grams.
   b) The molar mass of an element divided by Avogadro's number gives the mass of one atom of that element in grams.

2. If you have 1 g samples of several different elements, will the sample with the largest or smallest molar mass contain the fewest atoms? Explain.

3. Write the units that result from the following mathematical operations.
   a) number of objects / Avogadro's number =
   b) moles × Avogadro's number =
   c) mass / molar mass =
   d) moles × molar mass =

Problems

1. The atomic mass of ^{35}Cl is 34.971 amu and the atomic mass of ^{37}Cl is 36.970 amu. In a natural sample, 75.77% of the atoms are ^{35}Cl, and 24.23% are ^{37}Cl. Describe how you can calculate the molar mass of chlorine from these data, then calculate a value for the molar mass of chlorine.

2. A mass of 32.0 g of oxygen reacts completely with 6.02214×10^{23} atoms of carbon.

   a) What is the ratio of moles of C to moles of O in the product?

   b) Given that mass is conserved in a chemical reaction, what is the mass of the product produced?

   c) Is the product carbon monoxide, CO, or carbon dioxide, CO_2?

ACTIVITY 03-4

Determination of Molecular Formulas

WHY?

Molecular formulas are called chemical formulas. They tell you how many atoms of each element are present in a molecule. Molecular formulas can be determined by measuring the mass of each element present in a sample of the compound and the molar mass. This conversion of macroscopic quantities of material (grams) to the microscopic composition (number of atoms of each element present in a molecule) is used by chemists, biochemists, pharmacologists and others who work in the production of new materials for research, technology, and medicine.

LEARNING OBJECTIVE

- Understand the relationship between the mass percent composition of a chemical compound, its molar mass, and its molecular formula

SUCCESS CRITERIA

- Quickly calculate mass percent composition from a molecular formula
- Determine the empirical formula of a compound from its mass percent composition
- Determine the molecular formula of a compound from its empirical formula and molar mass

PREREQUISITES

- **Activity 03-1:** *Molecular Representations*
- **Activity 03-3:** *Mole and Molar Mass*

INFORMATION

The molecular or chemical formula of a molecule tells you the number of atoms of each element that comprise the molecule. It also tells you the number of moles of each element needed to make 1 mole of the compound. For example, the molecular formula for glucose (sugar), $C_6H_{12}O_6$, tells you that each molecule contains 6 carbon atoms, 12 hydrogen atoms, and 6 oxygen atoms, and that the corresponding number of moles of each element is needed to make 1 mole of glucose.

The percent composition by mass of a compound provides information that is useful in determining molecular formulas. The mass percent of an element in a sample of a pure compound is calculated in the following way:

$$\text{mass \% of an element} = \frac{\text{mass of the element present in a sample}}{\text{total mass of the sample}} \times 100\%$$

The mass % of an element in a compound can also be calculated if the molecular formula is known. This calculation is done in the following way:

$$\text{mass \% of an element} = \frac{\text{mass of the element in one mole of the compound}}{\text{molar mass of the compound}} \times 100\%$$

For example, to calculate the percent carbon in propene, C_3H_6: the mass of carbon in 1 mol of propene is 3 × 12.011 g/mol = 36.033 g/mol. The molar mass of propene is 42.080 g/mol, so dividing 36.033 by 42.080 and multiplying by 100% produces 85.63% carbon.

In addition to representation with a molecular formula, molecules can also be represented by a line structure as illustrated in **Model 1** below, and as first introduced in Activity 03-1, Molecular Representations. In a line structure, carbon atoms are at the end of each line and at the points where two lines come together. The lines represent bonds that hold the atoms together. A single line represents a single bond, a double line represents a stronger double bond, and a triple line represents an even stronger triple bond. Hydrogen atoms are not explicitly included in a line structure. Bonds between carbon and hydrogen atoms are implicit in a line structure because carbon atoms in a molecule must have four bonds. So the missing bonds in a line structure are the bonds between carbon and hydrogen.

MODEL 1: WHAT DO THESE DIFFERENT COMPOUNDS HAVE IN COMMON?

Name	Line Structure	Molecular Formula	Mass % Composition	
			C	H
ethene				
propene	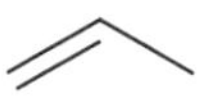	C_3H_6	85.63	14.37
1-butene	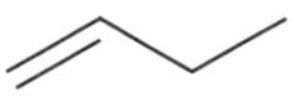			
2-butene	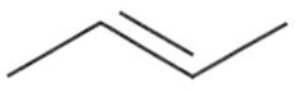			
1-pentene				
2-pentene	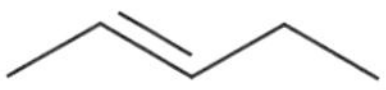			

TASK

Complete the table in **Model 1** by filling the missing molecular formula and mass percent composition information.

KEY QUESTIONS

1. What are common features of the 6 compounds in **Model 1** that you can identify from the table?

2. Is it possible to determine the molecular formula of a compound from the mass percent composition? Explain why or why not.

3. How are the molecular formulas of the six compounds in **Model 1** similar, and how are they different?

INFORMATION

An *empirical formula* for a compound contains integer subscripts that provide the ratio of elements in the compound by using the smallest whole numbers. The molecular formula is some multiple of the empirical formula. Additional information, for example the molar mass, is needed to determine the molecular formula from the empirical formula.

KEY QUESTIONS

4. What is the empirical formula for the compounds in **Model 1**?

5. How are the molecular formulas for the compounds in **Model 1** related to the empirical formulas of these compounds?

MODEL 2: CHEMICAL ANALYSIS OF ACETIC ACID

Acetic acid is the active ingredient in vinegar. A chemical analysis of 157.5 g of acetic acid provided the following information:

Element	Mass of Element (g)	Moles of Element	Whole Number Ratio
carbon	63.00	5.246	5.246 / 5.246 = 1
oxygen	83.93	5.246	5.246 / 5.246 = 1
hydrogen	10.57	10.486	10.486 / 5.246 = 2

moles = mass of element / molar mass of element

The 1:1:2 ratio in the last column means that the empirical formula is (COH_2).

KEY QUESTIONS

6. How was the number of moles of each element in the acetic acid sample calculated?

mass of element / molar mass of element

7. How were the ratios of the elements in acetic acid determined from the moles of each element present in the sample? there are 2 H so ratio for it is 2

8. How is the empirical formula determined from the ratios of the elements present in the sample?

9. What information does the empirical formula provide?

simplified ratio

10. What is the relationship between the ratio of moles of each element present in the sample and the ratio of the number of atoms of each element present in each molecule of acetic acid?

11. Why is the atomic mass percent composition of an unknown chemical compound an important quantity to determine in a chemical analysis?

EXERCISES

1. Sodium carbonate has the molecular formula Na_2CO_3. Calculate the molar mass of this compound, the mass percent composition of each element in the compound, and the mass of each element present in a 73.6 g sample.

Na = 22.989
C = 12.01
O = 15.99

molar mass = 105.958

$\frac{2(22.989)}{105.958} \times 100 = 43.4\%$ (.434 × 73.6) = 31.9

$\frac{12.01}{105.958} \times 100 = 11.3\%$ (.113 × 73.6) = 8.3

$\frac{3(15.99)}{105.958} \times 100 = 45.3\%$ (.453 × 73.6) = 33.3

2. A sample of baking soda was found to consist of 9.122 g Na, 0.4000 g H, 4.766 g C, and 19.04 g O. Calculate the moles of each element present, and determine the empirical formula of baking soda.

$9.122\,g\,Na\left(\frac{1\,mole}{22.989\,g}\right) = .397\,mole$ $\qquad Na_{.397}H_{.4}C_{.397}O_{1.19}$

$.4\,g\,H\left(\frac{1\,mole}{1\,g}\right) = .4\,mole$ $\qquad \frac{.397}{.397} = 1$

$4.766\,g\,C\left(\frac{1\,mole}{12.011\,g}\right) = .397\,mole$ $\qquad NaHCO_3$

$19.04\,g\,O\left(\frac{1\,mole}{15.999}\right) = 1.19\,mole$ $\qquad \frac{1.19}{.397} = 2.99$

PROBLEMS

1. Show how you can determine the molecular formula for acetic acid from the empirical formula CH_2O and the added information that the molar mass of acetic acid is 60.05 g/mole.

$12.011 + 2(1.007) + (15.99) = 30.015$

$\frac{60.05}{30.015} = 2 \qquad 2(CH_2O) = C_2H_4O_2$

2. One of the chlorofluorocarbons (a freon), which is used in refrigerator compressors and contributes to destruction of ozone in the upper atmosphere, has a molar mass of 132.9 g/mole and a percent composition of 53.34% Cl, 28.59% F, and 18.07% C. Hint: Whenever the data give the percent of the elements present rather than the mass, you can assume any sample size and calculate the mass. Why is it particularly convenient to assume the data came from a 100 g sample?

   a) How many grams of chlorine are there in a 100 g sample of the freon?

   ~~28.59~~ 53.34 $\frac{1}{35.4}$ = 1.55 mol   53.34g

   b) How many moles of chlorine are there in a 100 g sample of the freon?

   c) What is the ratio of the moles of chlorine to the moles of carbon in this freon?

   d) Determine the empirical and molecular formulas of this freon.

3. Combustion of 10.68 mg of a compound containing only C, H, and O produces 16.01 mg CO_2 and 4.37 mg H_2O. The molar mass of the compound is 176.1 g/mol. What is the empirical formula of the compound? What is the molecular formula of the compound?

$16.01 \text{ mg } CO_2 + 4.37 \text{ mg } H_2O \rightarrow C\ H\ O + O_2$

$16.01 \text{ mg } CO_2 \left(\frac{12 \text{ mg C}}{44 \text{ mg } CO_2} \right) = 4.37 \text{ mg C}$

$4.37 \text{ mg } H_2O \left(\frac{2 \text{ mg H}}{18 \text{ mg } H_2O} \right) = 0.485 \text{ mg H}$

$10.68 - 4.37 - 0.485 = 5.82$

$4.37 \text{ mg C} \left(\frac{1 \text{ g}}{1000 \text{ mg}} \right) \left(\frac{1 \text{ mol}}{12 \text{ g}} \right) = 0.000364 \text{ moles C}$

.364 mmoles

0.485 mmoles H

0.364 mmoles O

$C_{.364} H_{0.485} O_{.364} = CH_{1.33}O$

$3 \times 1.33 = 4 \quad C_3H_4O_3$

HINTS The empirical formula gives the proportions in terms of mole ratios of each element present in the compound. To find the empirical formula, you first need to determine how many moles of each element are present in a sample of the compound.

1. You can determine the mass and moles of carbon in the compound from the mass of carbon dioxide that was produced in the combustion. Can you explain why the fraction of the carbon dioxide mass that is due to carbon is 12.01 / 44.01?

2. You can determine the mass and moles of hydrogen in the compound from the mass of water that was produced in the combustion by first determining the fraction of the mass of water that is due to hydrogen.

3. You can determine the mass and moles of oxygen because the sum of O, C, and H masses must add up to 10.68 mg, which is the mass of the sample.

4. From the moles of C, H, and O, you can determine the mole ratios, which leads to the empirical formula.

5. Given the empirical formula and the molar mass of 176.1 g/mol, you can identify the molecular formula.

Activity 04-1

Balanced Chemical Reaction Equations

Why?

Chemical reaction equations are fundamental tools for communicating how chemical compounds are synthesized and changed. Representing a chemical reaction with an equation is one key to understanding chemical change. To make use of these equations, you need to be able to write formulas for chemical compounds, balance the chemical equation, and deduce information from it about the amounts of material needed for the reaction and the amounts of material produced by the reaction.

Learning Objective

- Understand and make use of chemical reaction equations

Success Criteria

- Write and balance chemical reaction equations
- Determine the amounts of substances consumed and produced in a chemical reaction

Prerequisites

- **Activity 03-1:** *Molecular Representations*
- **Activity 03-3:** *Mole and Molar Mass*

Model: Burn, Propane, Burn! (A One-Act Play in Two Scenes)

(Have fun by acting this play out in class! See if you can extract the key ideas about balancing reaction equations.)

Scene 1 opens with Nimka, Chris, and Diana hanging out and talking about their favorite subject–chemistry.

Nimka:	I just read that when propane burns in oxygen, carbon dioxide and water are produced.
Chris:	I know the molecular formulas, so I can write an equation for it.
Chris writes:	$C_3H_8 + O_2 \longrightarrow CO_2 + H_2O$
Nimka:	That's neat! Does it mean that one molecule of propane combines with one molecule of oxygen as the reactants?
Chris:	Yup! To produce the products: molecules of carbon dioxide and water.
Diana:	What happened to all the atoms?
Chris:	What do you mean?
Diana:	Well, you know, there are 13 atoms on the left side and only 6 atoms on the right side.
Nimka:	It's OK, there are 2 molecules on the left and 2 molecules on the right, and Dr. Dave told us that reaction equations need to be balanced.
Diana:	But molecules are just atoms connected together, and the atoms don't balance.
Nimka:	Yeah, on the right side there are too many oxygen atoms and too few carbon and hydrogen atoms.

Burn, Propane, Burn! (continued)

Chris: Perhaps the protons, neutrons, and electrons rearranged, and carbon and hydrogen turned into oxygen.

(The three friends huddle together to count the number of protons, neutrons, and electrons comprising the reactants and products.)

Chris: Darn, that doesn't seem to work.

Diana: Maybe if we add more molecules.

Chris: Of what?

Diana: Carbon dioxide and water.

Nimka: And then put more oxygen on the left!

(The three friends huddle together again and come up with a new equation.)

Chris: We got it! Look, it works!

Chris writes: $C_3H_8 + 5\ O_2 \longrightarrow 3\ CO_2 + 4\ H_2O$

Nimka: Let's go ask Dr. Dave if this is right.

Scene 2 opens in Dr. Dave's office.

Dr. Dave: Yes, you three figured that out. The number of atoms of each element must be the same on both sides of the equation because atoms are not created or destroyed in a chemical reaction, just regrouped to form different molecules. That's the key point! Let me say that a bit louder — ***That's the key point!***

Nimka (with emphasis on "that"): So **that** is what is meant by balanced.

Dr. Dave: Well almost. Here's something else for you to think about. Why isn't the following reaction equation balanced? $Cu^{2+} + Ag \longrightarrow Cu + Ag^{+}$

Nimka: It looks balanced, 1 copper and 1 silver on each side!

Diana: But what about the charge? On the left it is +2, and on the right it is +1.

Dr. Dave: That's right. An electron got transferred from silver to copper in the reaction, but two electrons are needed to reduce copper from +2 to 0.

Nimka: So you mean we need 2 silver atoms to provide 2 electrons?

Dr. Dave: You got it! Congratulations! There are a couple of other points about writing balanced reaction equations. Would you like to hear about them?

The three friends shout loudly in unison: "Yes!"

Dr Dave: A complete equation specifies the state of the reactants and products with a symbol: (s) for solid, (g) for gas, (l) for liquid, and (aq) for dissolved in water. So the copper/silver reaction is better written as,

$$Cu^{2+}(aq) + 2Ag(s) \longrightarrow Cu(s) + 2Ag^{+}(aq)$$

Also, a balanced reaction equation can be read two ways: in terms of molecules reacting or in terms of moles reacting. For example, you could say that 1 molecule of propane reacts with 5 molecules of oxygen to produce 3 molecules of carbon dioxide and 4 molecules of water. Or you could say, 1 mole of propane reacts with 5 moles of oxygen to produce 3 moles of carbon dioxide and 4 moles of water.

(The scene closes as Dr. Dave rushes off to a faculty meeting, and the three friends race to their favorite class — chemistry recitation where they get to work in teams, talk about chemistry, and figure out how to ace the exam by understanding rather than memorizing.)

Key Questions

1. What are the reactants in the combustion of propane?

2. What are the products in the combustion of propane?

3. What meaning is given to the arrow in a chemical reaction equation?

4. In the reaction equation, what information is provided by the numerical subscripts in the molecular formulas for the reactants and products?

5. In a balanced reaction equation, what information is provided by the stoichiometric coefficients? Note: The number in front of a reactant or product is called a *stoichiometric coefficient.* The absence of an explicit stoichiometric coefficient means the value is 1.

6. How does the second reaction equation that Chris wrote differ from the first reaction equation that he wrote?

7. Why is the second reaction equation that Chris wrote more useful than the first reaction equation that he wrote?

8. What two things must be true for a reaction equation to be balanced?

9. Why is it possible to interpret a reaction equation in each of the following two ways?
   a) As specifying how many molecules of reactants are consumed to produce the specified number of product molecules.
   b) As specifying how many moles of reactants are consumed in producing the indicated number of moles of products.

EXERCISES

1. Write balanced reaction equations for the following reactions. Be sure to add the states if they are missing.
   a) $Zn + Ag^+ \longrightarrow Zn^{2+} + Ag$
   b) $Na_2CO_3(s) + HCl(aq) \longrightarrow NaCl(aq) + CO_2(g) + H_2O(l)$
   c) $CH_3OH(l) + O_2 \longrightarrow CO_2(g) + H_2O(g)$
   d) In the gas phase: $N_2 + H_2 \longrightarrow NH_3$
   e) $LiOH(s) + CO_2 \longrightarrow Li_2CO_3(s) + H_2O(l)$
   f) $Pb(NO_3)_2\ (aq) + H_2S(g) \longrightarrow PbS(s) + HNO_3(aq)$

2. Using the balanced reaction equation for the combustion of propane, determine the number of moles of oxygen that would react with 0.50 mol propane and the number of moles of carbon dioxide that would be produced.

$C_3H_8 + 5O_2 \rightarrow 3CO_2 + 4H_2O$

~~.5 C_3H_8 (1 mole / 44 C_3H_8)~~ (5 /)

.5 (5 O_2 mole / 1 mole) = 2.5 O_2

.5 (3 CO_2 mole / 1 mole) = 1.5 CO_2

3. Using the reaction equation for the combustion of propane, determine the number of grams of oxygen that would react with 44 g of propane and the number of grams of water that would be produced.

44 g C_3H_8 (1 mole / 44 C_3H_8) (5 O_2 mole / 1 mole) (32 g / 1 mole)

Activity 04-2

Dissociation and Precipitation Reactions

Why?

Some compounds dissolve and dissociate into ions when added to water. Other compounds are not soluble in water and form a precipitate when solutions of soluble compounds are mixed. Compounds dissolving in water and precipitating from aqueous solutions are associated with health issues, environmental issues, and manufacturing processes. In order to use chemical reactions or to evaluate the effects of chemical reactions, you need to be able to identify the type of reaction, the conditions when it can occur, and describe it by a balanced reaction equation.

Learning Objective

- Recognize the characteristics of dissociation and precipitation

Success Criteria

- Write and balance ionic, net ionic, and molecular reaction equations
- Identify whether precipitation will occur when solutions are mixed

Prerequisites

- **Activity 03-1:** *Molecular Representations*
- **Activity 04-1:** *Balanced Chemical Reaction Equations*

Model: Dissociation and Precipitation

Many compounds dissolve and dissociate into ions when added to water, e.g.

$$Na_2S(s) \longrightarrow 2Na^+(aq) + S^{2-}(aq)$$

$$Cd(NO_3)_2(s) \longrightarrow Cd^{2+}(aq) + 2NO_3^-(aq)$$

$$NaNO_3(s) \longrightarrow Na^+(aq) + NO_3^-(aq)$$

Other compounds are insoluble in water, e.g. cadmium sulfide. So when solutions of sodium sulfide and cadmium nitrate are mixed together, solid cadmium sulfide forms. This solid is called a *precipitate*.

$$2Na^+(aq) + S^{2-}(aq) + Cd^{2+}(aq) + 2NO_3^-(aq) \longrightarrow CdS(s) + 2Na^+(aq) + 2NO_3^-(aq)$$

This precipitation reaction equation is called the *ionic reaction equation* because it includes all the ions in the solutions that were mixed together.

The precipitation reaction can also be written as a *net ionic equation* where only the ions that participate in the precipitation reaction are included. The ions that are left out and don't participate are called *spectator ions*.

$$Cd^{2+}(aq) + S^{2-}(aq) \longrightarrow CdS(s)$$

The reaction can also be written as a *balanced molecular equation*, where the (aq) designation means the species as it exists in aqueous solution, which in this case would be the ion dissociation products.

$$Na_2S(aq) + Cd(NO_3)_2(aq) \longrightarrow CdS(s) + 2NaNO_3(aq)$$

Suggestions contributed by Vicky Minderhout, Seattle University

Key Questions

1. What is the characteristic identifying feature of a dissociation reaction?

2. What is the characteristic identifying feature of a precipitation reaction?

3. What is common to the ionic reaction equation, the net ionic equation, and the molecular equation?

4. What is the characteristic identifying feature of the ionic reaction equation, the net ionic equation, and the molecular equation?

Exercises

1. Use your textbook or other reference resource to find solubility rules for ionic compounds. Ionic compounds are those compounds that dissolve and dissociate to produce ions in solution or precipitate from ions in solution because they are insoluble.

   a) Use atomic symbols to list those cations that generally form soluble ionic compounds.

   Li^+ Na^+ K^+ NH_4^+

   b) Use atomic symbols to list those anions that generally form soluble ionic compounds.

   NO_3^- $C_2H_3O_2^-$ ClO_4^-

   Cl^- Br^- I^-

c) Use atomic symbols to list those anions that generally form insoluble ionic compounds.

Most sulfate salts are soluble

d) In your lists of a) through c), identify any exceptions to the general rule.

Ag^+ Pb^{2+} Hg_2^{2+}

2. For each of the pairs of reactants in a) through c) below,

   i) Write the formulas for the reactants.

   ii) Below these formulas write the ions that these reactants produce through dissociation in aqueous solution.

   iii) Predict whether a precipitation reaction will occur. Consider the possible compounds produced when the anions and cations of the reactants exchange partners. If any of these is predicted to be insoluble from your solubility rules, then the reaction is a precipitation reaction.

   iv) Write the balanced overall ionic reaction using the formulas for the reactants and products. Be sure to include the (states) for each.

   v) Write the balanced net ionic equation, omitting the spectator ions.

   vi) Write the balanced molecular equation.

   a) lead(II) acetate and sodium carbonate in water

   i) $Pb^{2+}(C_2H_3O_2)_2^- + Na^+CO_3 \rightarrow Pb^{2+}CO_3\,(s) + 2Na\,C_2H_3O_2\,(aq)$

   ii) ______

   iii) ______

   iv) ______

   v) ______

   vi) ______

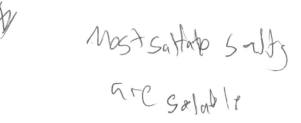

b) calcium chloride and sodium phosphate in water

i) $CaCl_2$ + $NaPO_4$

ii) Ca

iii)

iv)

v)

vi)

c) ammonium chloride and copper(II) sulfate in water.

i)

ii)

iii)

iv)

v)

vi)

Activity 04-3

Introduction to Acid – Base Reactions

Why?

Chemical reactivity can be understood by identifying the reactants as acids or bases. Blood acidity is important for health, acid-base reactions are used in industry to produce many materials, and acid rain is a complex environmental issue. You need to recognize acids and bases and evaluate their effects in order to use these reactions for benefit and to avoid harm.

Learning Objective

- Identify the characteristics of acids and bases

Success Criteria

- Quickly recognize acids and bases
- Write acid-base reactions correctly

Prerequisites

- **Activity 03-1:** *Molecular Representations*
- **Activity 04-1:** *Balanced Chemical Reaction Equations*

Information

An *acid* is a chemical species that donates a proton to another species.

A *base* is a chemical species that accepts a proton from another species.

Task

Label each of the reactants in the Model below as *acid* or *base*, using the definitions of an acid and a base given previously.

Model: Acids and Bases

1. $HCl(aq) + NH_3(aq) \longrightarrow NH_4^+(aq) + Cl^-(aq)$

2. $H_2O(l) + NH_3(aq) \longrightarrow NH_4^+(aq) + OH^-(aq)$

3. $HCl(aq) + H_2O(l) \longrightarrow H_3O^+(aq) + Cl^-(aq)$

4. $H_3PO_4(aq) + H_2O(l) \longrightarrow H_2PO_4^-(aq) + H_3O^+(aq)$

5. $H_2PO_4^-(aq) + H_2O(l) \longrightarrow HPO_4^{2-}(aq) + H_3O^+(aq)$

aq = aqueous
l = liquid
s = solid

KEY QUESTIONS

1. For the reactions in the Model, how did you determine whether a reactant was an acid or a base?

2. Does any substance in the Model act both a proton donor (acid) and as a proton acceptor (base)? Explain how this can or cannot occur. A substance that can act either as an acid or a base is called amphoteric; an amphoteric substance that can either donate or accept a proton is also called amphiprotic.

3. Would you call sodium hydroxide, NaOH, an acid or a base? Why? Sodium hydroxide ionizes in water to give Na^+(aq) and OH^-(aq).

4. What is an example of an acid or a base that you have encountered outside the chemistry laboratory?

EXERCISES

1. In the context of reversing the reactions in the Model, label the acids on the right-hand side of the reaction equations in the Model.

2. In the context of reversing the reactions in the Model, label the bases on the right-hand side of the reaction equations in the Model.

3. Write the acid-base reaction for NH_3 reacting with HNO_2 and identify the acid and the base on the reactant side and on the product side.

4. Complete the following table for the reaction equations in the Model. Label each reactant and product as *acid* or *base*. Reaction 1 is done for you.

Reaction	Reactant	Corresponding Product
1	HCl (acid)	Cl^- (base)
1	NH_3 (base)	NH_4^+ (acid)
2	H_2O (acid)	NH_4 (base)
2	NH_3 (base)	OH (acid)
3	HCl (acid)	H_2O (base)
3	H_2O (base)	Cl (acid)
4		
4		
5		
5		

5. Examine your entries in the table in Exercise 4 and identify how the acids and the bases in each row differ.

Information

Notice in the Model that, following the loss of a proton, an acid on the left-hand side of a reaction equation produces a corresponding base on the right-hand side. These acid-base pairs are called *conjugate acid-base pairs*. A *conjugate acid-base pair* is any two substances that can be converted from one to the other simply by transferring a proton between them.

Exercises

6. Complete the following table of conjugate acid-base pairs.

	Acid	Base
1		CN^-
2	HNO_3	
3	H_2CO_3 (a diprotic acid)	
4		CO_3^{2-}
5		NH_2^-
6	$HClO_4$	
7		CH_3NH_2
8	$(CH_3)_2NH_2^+$	

7. Draw the structural formula of CH_3NH_2, and add 2 dots to the nitrogen atom to represent the pair of nonbonding electrons. Use your drawing to explain why the hydrogen ion attaches to nitrogen and not to carbon.

8. For the following reaction, identify the reactant that is an acid, the reactant that is a base, and the two conjugate acid-base pairs present.

$$H_2SO_4 + HPO_4^{2-} \longrightarrow HSO_4^- + H_2PO_4^-$$

Activity 04-4

Electron Transfer Reactions

Why?

Corrosion of metals, combustion of fuels, the generation of electricity from batteries, and many biological processes involve electron transfer reactions. In these reactions, which are also called oxidation-reduction or redox reactions, electrons are transferred from one chemical species to another. By understanding the principles of redox reactions, scientists and engineers can prevent corrosion, design conditions for more efficient combustion, produce new kinds of batteries, and increase the lifespan of materials and biological systems.

Learning Objectives

- Distinguish electron transfer reactions from other kinds of reactions
- Characterize electron transfer reactions in terms of electron flow and changes in oxidation numbers

Success Criteria

- Assign oxidation numbers to atoms in a chemical compound
- Identify the oxidizing agent and the reducing agent in a redox reaction
- Identify the number of electrons transferred in a redox reacton

Prerequisites

- **Activity 02-3:** *Periodic Table of the Elements*
- **Activity 03-1:** *Molecular Representations*
- **Activity 04-1:** *Balanced Chemical Reaction Equations*

Information

Electronegativity is a measure of the ability of an atom to attract electrons in a molecule. Electrons are pulled toward the atoms that have the larger electronegativities. A chart of electronegativity values is provided on the back cover of this book.

The *oxidation number* of an atom is the charge an atom would have if each of its valence electrons were assigned to the more electronegative atom in each bond. The oxidation number helps you keep track of electron flow in redox reactions even though electrons may not be transferred completely. The oxidation number of an atom in a chemical compound is the basic tool used to identify and understand redox reactions.

When an atom is *oxidized*, it loses electrons, and its oxidation number increases. When an atom is *reduced*, it gains electrons, and its oxidation number decreases. So in forming CO, C has been oxidized, and O reduced because oxygen has a higher electronegativity than carbon.

The reactant containing the element being oxidized is called the *reducing agent* because it is providing electrons to another atom. The reactant containing the atom being reduced is called the *oxidizing agent* because it is removing electrons from another atom.

To summarize: *Oxidation* is the loss of electrons from a chemical species. *Reduction* is the gain of electrons by a chemical species. To remember these definitions, think *reduction* means the *oxidation number* is *reduced*.

Model 1: Guidelines for Assigning Oxidation Numbers

The key to understanding oxidation numbers is to remember that the *oxidation number* is defined as the charge an atom would have if each of its valence or bonding electrons were assigned to the more electronegative atom in each of its bonds. The more electronegative atom thereby ends up with a complete valence octet of electrons. The rationale below is based on this idea.

Chemical Species	Oxidation Number	Rationale
O_2 Fe metal Na metal	0 0 0	The oxidation number of an atom of a pure element is always 0 because there is no electronegativity difference when all the atoms are the same.
Na^+	+1	In compounds, the oxidation number of the Group IA alkali metals is always +1 because these metals have one valence electron and have small electronegativity values.
Mg^{2+}	+2	In compounds, the oxidation number of the Group IIA alkaline earth metals is always +2 because these metals have two valence electrons and have small electronegativity values.
F^-	−1	In compounds, the oxidation number of fluorine is always −1 because it is the most electronegative element and always attracts one electron from another atom.
monatomic ion		The oxidation number of a monatomic ion is its charge.
polyatomic ion		For a polyatomic ion, the sum of the oxidation numbers of the atoms equals the charge on the ion.
neutral molecule		For a neutral molecule, the sum of the oxidation numbers of all the elements is zero.
H_2O, CH_4	H = +1	In compounds with more electronegative elements, the oxidation number of H is +1.
LiH, NaH	H = −1	When H is combined with metals having a smaller electronegativity, its oxidation number is −1.
H_2O, NO_3^-	O = −2	The oxidation number of O is generally −2.
H_2O_2, Na_2O_2	O = −1	However in peroxides, like hydrogen and sodium peroxide, its oxidation number is −1 because the O atoms can't attract electrons from each other.
CCl_4, KCl	Cl = −1	Group VIIA elements (halogens) generally are assigned oxidation numbers of −1 in compounds with other elements because the halogen has the larger electronegativity.
$HClO_4$	Cl = +7	In cases where the halogen has the smaller electronegativity, e.g., when combined with oxygen or other halides, the oxidation number can be positive. For example, in $HClO_4$, oxygen is the most electronegative element so each oxygen is assigned both electrons in the bonds leaving H and Cl with no valence electrons and oxidation numbers of +1 and +7 respectively.
ClF_5	Cl = +5	F is the more electronegative element in ClF_5 so each F is assigned both electrons in the bonds leaving Cl with 2 electrons and a +5 oxidation number.

Key Questions

1. What is the oxidation number of a pure element such as H_2, N_2, K, Ni, S_8, or Cu?

   ∅

2. For the polyatomic ion SO_4^{2-}, what is the sum of the oxidation numbers of sulfur and the four oxygen atoms?

   0

3. In terms of relative electronegativities, why is the oxidation number of hydrogen +1 in compounds with C, N, O, and Cl, but −1 in compounds with metals like Li, Na, and Al?

4. In terms of relative electronegativities, why does chlorine have a negative oxidation number in CCl_4, but a positive oxidation number in HClO?

Got It!

Explain why you agree or disagree with each of the following statements in terms of the number of electrons and how they are assigned to the atoms.

1. The most electronegative atom in a compound generally will have a negative oxidation number equal to the number of electrons needed to complete its valence octet.

2. The least electronegative atom in a compound will have a positive oxidation number.

3. The sum of all oxidation numbers must equal the charge on the species. For a neutral compound the sum must be 0, for a cation like NH_4^+ the sum must be +1, and for an anion like SO_4^{2-}, the sum must be −2.

EXERCISES

1. Using the guidelines in the model and the insight gained from your answers to the *Key Questions* and *Got It!* section, assign oxidation numbers to all atoms in the following compounds or ions.

O_2	H_2O	CH_4	CO_2	SF_6
0̸	H+1 O-2	H+1 C-4	C+4 O-2	S -3 F+1
NaNO₃ H+1 -2	SO_4^{2-}	H_2O_2	$CuCl_2$	$HClO_2$
+1 +5 -2	+6 -2	+1 -1	+2 -1	
KMnO₄	HCN	H₂S	P₄	Na₂SO₃
	K₂CrO₄	Cr₂O₃	Na₂Cr₂O₇	

MODEL 2: ELECTRON TRANSFER REACTIONS

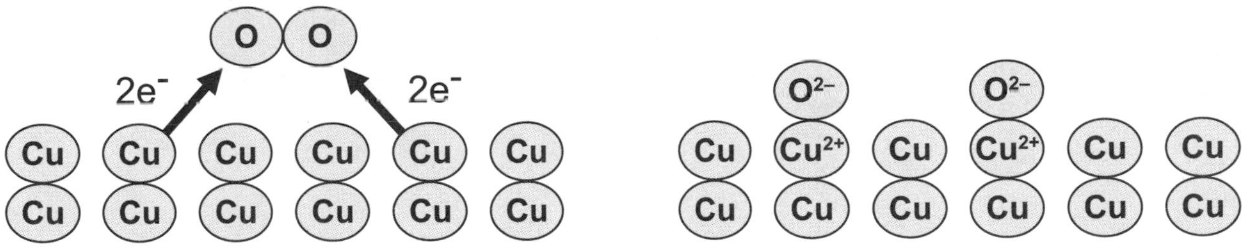

In the oxidation of copper metal, electrons are transferred from copper atoms to molecular oxygen to form copper oxide. The chemical reaction equation is

$$2Cu + O_2 \rightarrow 2CuO$$

A possible mechanism for this reaction is that an oxygen molecule adsorbs on the copper metal surface and dissociates. Then, two electrons are transferred from each of two copper atoms to the two oxygen atoms. As a result, the oxidation number of each copper atom changes from 0 to +2 because two electrons were lost, and the oxidation number of each oxygen atom changes from 0 to −2 because two electrons were gained.

Oxygen is reduced because the oxidation number decreases, and copper is oxidized because the oxidation number increases. Oxygen is therefore the oxidizing agent, and copper is the reducing agent.

You can recognize a redox reaction by determining whether oxidation numbers change or not.

KEY QUESTIONS

5. In the oxidation of copper, how many electrons are transferred from one copper atom to one oxygen atom?

6. What are the oxidation numbers of the atoms in Cu metal, O_2, and CuO?

7. In the oxidation of copper, what is the chemical species that is oxidized, and what is the chemical species that is reduced?

8. In the oxidation of copper, what is the oxidizing agent, and what is the reducing agent?

9. If some chemical species contains an atom that has an increase in the oxidation number during a chemical reaction, does that species act as the oxidizing agent or the reducing agent?

10. How can you distinguish a redox reaction from other types of chemical reactions?

EXERCISES

2. Assign oxidation numbers to all the atoms in the following reactions. For the redox reactions, identify the atom oxidized, the atom reduced, the oxidizing agent, and the reducing agent.

   a) This reaction is used to produce iron from iron ore.

$$FeO(s) + CO(g) \longrightarrow Fe(s) + CO_2(g)$$

$Fe\,O^{-2} + C^{+4}O^{-2} \rightarrow Fe^{0} + C^{+}O_2^{-2}$

Redox (yes/no)	Atom Oxidized	Atom Reduced	Oxidizing Agent	Reducing Agent
yes	~~Fe~~ C	Fe	FeO	CO

b) This reaction is used to produce industrially important quicklime from limestone.

$$\overset{+2\ +4\ -2}{CaCO_3(s)} \longrightarrow \overset{+2\ -2}{CaO(s)} + \overset{+4\ -2}{CO_2}$$

$$\overset{+2}{Ca}\overset{+4}{C}\overset{-2}{O_3} \rightarrow \overset{+2}{Ca}\overset{-2}{O} + \overset{+4}{C}\overset{-2}{O_2}$$

Redox (yes/no)	Atom Oxidized	Atom Reduced	Oxidizing Agent	Reducing Agent
No				

c) This reaction can be used to produce hydrogen.

$$2H_3O^+(aq) + Mg(s) \longrightarrow Mg^{2+}(aq) + H_2(g) + 2H_2O(aq)$$

Redox (yes/no)	Atom Oxidized	Atom Reduced	Oxidizing Agent	Reducing Agent
yes	Mg	H		

d) This reaction is so fundamental that you should recognize it immediately.

$$HCl(aq) + NaOH(aq) \longrightarrow NaCl(aq) + H_2O(aq)$$

Redox (yes/no)	Atom Oxidized	Atom Reduced	Oxidizing Agent	Reducing Agent
No				

e) This reaction occurs in an automobile battery.

$$Pb(s) + 2HSO_4^-(aq) + PbO_2(s) + 2\,H_3O^+(aq) \longrightarrow 2\,PbSO_4(s) + 4\,H_2O(l)$$

Redox (yes/no)	Atom Oxidized	Atom Reduced	Oxidizing Agent	Reducing Agent
yes	Pb	Pb	PbO_2	Pb

f) This reaction causes sodium to burst into flame when tossed onto water.

$$Na(s) + 2\,H_2O(l) \longrightarrow 2\,NaOH(aq) + H_2(g)$$

Redox (yes/no)	Atom Oxidized	Atom Reduced	Oxidizing Agent	Reducing Agent
yes	Na	H	H_2O	Na

Activity 05-1

Limiting Reactants

Why?

Reactants are not always present in the exact amounts required by a balanced chemical reaction equation. In planning any cost-effective production process, it is necessary to recognize which component limits the amount of material that can be produced. Identifying the limiting reactant in a chemical reaction will strengthen your skills in dealing with moles, solution concentrations, and reaction stoichiometry. *Stoichiometry* means measuring amounts of material.

Learning Objectives

- Determine the amounts of material involved in a reaction
- Identify reactants that limit the extent of a reaction and the amount of product produced

Success Criteria

- Quick identification of the limiting and excess reactants
- Accuracy of calculations of the amounts of material reacting and being produced in chemical reactions

Prerequisites

- **Activity 03-1:** *Molecular Representations*
- **Activity 03-3:** *Mole and Molar Mass*
- **Activity 03-4:** *Determination of Molecular Formulas*
- **Activity 04-1:** *Balanced Chemical Reaction Equations*

Model 1: Limiting Ingredient

A cake recipe calls for:

2 cups of water
8 squares of chocolate
8 oz of butter
4 cups of flour
4 cups of sugar
4 eggs

Ingredients on hand:

lots of water

5 cups flour

4 cups sugar

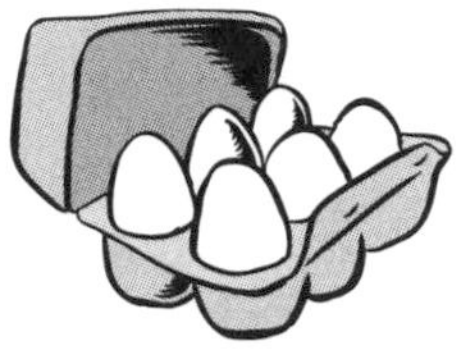

6 eggs

12 squares chocolate

16 oz butter

KEY QUESTIONS

According to **Model 1**, how much of each ingredient is necessary to make a cake?

water	flour	chocolate	sugar	butter	eggs

If you follow the recipe, using only the ingredients on hand in the Model, how much of each ingredient will be left over after you have baked the cake?

water	flour	chocolate	sugar	butter	eggs

3. Which of the ingredients on hand were in excess for the recipe?

4. Which of the ingredients on hand were consumed completely in making the cake?

5. Which of the ingredients limit or prevent you from making a second, smaller cake?

6. What would be a good definition for the term *limiting ingredient*?

7. What would be a good procedure or methodology to use for identifying the limiting component in some manufacturing process? Test your methodology by applying it to the following exercises.

EXERCISES

1. You want to make 10 dozen standard-sized cookies as specified by a recipe that requires 16 oz of butter, 4 eggs, 3 cups of flour, and 4 cups of sugar. In taking inventory of your supplies, you find that you have 16 oz of butter, 6 eggs, and 3 cups each of flour and sugar.

   a) Express the recipe for these cookies in the form of a reaction equation as started for you below.

   16 oz butter + 4 eggs + 3 cups flour, 4 cups sugar

   b) Which ingredient will limit the number of cookies you can make?

   sugar

   c) How many standard-sized cookies can you make?

2. You have 100 bolts, 150 nuts, and 150 washers. You assemble a nut/bolt/washer set using the following recipe or equation.

   2 washers + 1 bolt + 1 nut ⟶ 1 set

   a) How many sets can you assemble from your supply?

   b) Which is the limiting component? washers

3. This reaction of hydrogen with oxygen to produce water is described by the following recipe or reaction equation, which says that 2 molecules of hydrogen react with 1 molecule of oxygen to produce 2 molecules of water.

   $$2H_2 + O_2 \longrightarrow 2H_2O$$

   You react 150 H_2 molecules with 100 O_2 molecules to produce H_2O. Which is the limiting reactant, hydrogen or oxygen? How many water molecules can you produce from your supply of hydrogen and oxygen?

Model 2: Limiting Reactant in Moles

The reaction equation can be interpreted in terms of the number of molecules (see Exercise 3) or moles reacting (2 moles of hydrogen react with 1 mole of oxygen to produce 2 moles of water). The only difference is that while a mole can be divided into fractions of a mole, a molecule cannot be divided. Each square below represents one mole of hydrogen, oxygen, and water.

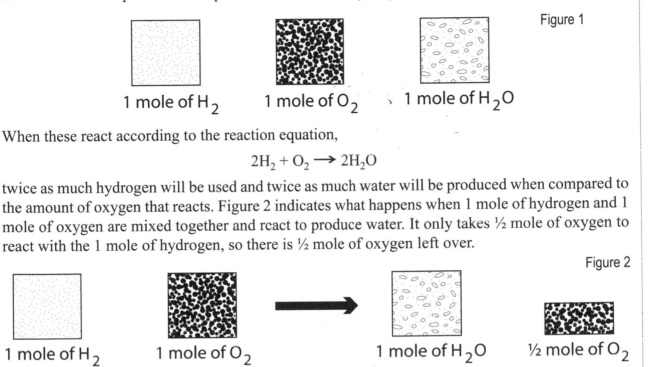

Figure 1

1 mole of H_2 1 mole of O_2 1 mole of H_2O

When these react according to the reaction equation,

$$2H_2 + O_2 \longrightarrow 2H_2O$$

twice as much hydrogen will be used and twice as much water will be produced when compared to the amount of oxygen that reacts. Figure 2 indicates what happens when 1 mole of hydrogen and 1 mole of oxygen are mixed together and react to produce water. It only takes ½ mole of oxygen to react with the 1 mole of hydrogen, so there is ½ mole of oxygen left over.

Figure 2

1 mole of H_2 1 mole of O_2 1 mole of H_2O ½ mole of O_2

Key Questions

8. How many moles of hydrogen reacted in the above diagram?
9. How many moles of oxygen remained after the reaction?
10. Was hydrogen or oxygen the limiting reactant?
11. In limiting reactant problems, you are sometimes given quantities of reactants in grams. Why do you need to convert these quantities to moles of reactants in order to solve the problem?

Exercises

4. If 6 moles of hydrogen (H_2) and 4 moles of oxygen (O_2) are mixed and reacted, which is the limiting reactant? How many moles of water would be produced?

H

5. If you had 1.73 moles of hydrogen (H_2) and 0.89 moles of oxygen (O_2), which would be the limiting reactant? How many moles of water can you produce from your supply of hydrogen and oxygen?

6. If you had 17.3 g of hydrogen and 8.91 g of oxygen, which is the limiting reactant, and how many grams of water could you produce?

Problems

1. Cisplatin is an antitumor agent. It has the molecular formula $Pt(NH_3)_2Cl_2$. How many grams of cisplatin can be produced if the limiting reactant is 1 kg of platinum?

2. Hydrogen cyanide is used in the production of cyanimid fertilizers. It is produced by the following reaction.

$$2\,CH_4 + 2\,NH_3 + 3\,O_2 \longrightarrow 2\,HCN + 6\,H_2O$$

a) How much hydrogen cyanide can be produced starting with 100 kg of each of the reactants?

b) Which is the limiting reactant?

Activity 05-2

Solution Concentration and Dilution

Why?

You can determine the number of molecules or moles present from the volume of a solution, provided you know the concentration of the solution. You will encounter problems and situations in this course, other courses, and many jobs, where you need to know the concentration of solutions. Laundry detergents, medicines, food products, cosmetics, steel, gasoline, and many other materials must contain specific amounts of certain chemical compounds to have desired properties.

Learning Objectives

- Understand and use the concept of molarity
- Determine how concentration changes upon dilution

Success Criteria

- Accurate calculations of solution concentration
- Accurate conversion of concentration into the amount (moles or mass) of material

Prerequisite

- **Activity 03-3:** *Mole and Molar Mass*

Information

A solution is formed when one substance is dissolved in another. The substance present in the larger amount is called the *solvent* and the other substance is called the *solute*. Water is a common solvent, and solutions involving water as the solvent are called *aqueous solutions* (abbreviated aq).

Model: Preparation of a Standard Solution

To prepare a solution of known concentration, 60.57 g (0.2059 mol) of potassium dichromate are transferred to a 500.0 mL volumetric flask. Some distilled water is added to the flask, the solid is dissolved, and additional distilled water is added to bring the water level to the 500.0 mL mark on the flask. The concentration of the potassium dichromate solution that was prepared is 0.4118 M.

molar concentration = molarity of the solution

molarity (M) = (moles of solute) / (liters of solution)

0.4118 M = 0.2059 mol / 0.5000 L

$[K_2Cr_2O_7]$ = 0.4118 M

Square brackets around the chemical formula represents the molar concentration of the substance.

Key Questions

1. In the model, what two factors determine the molar concentration of a solution?

2. What unit is used for solution concentration in the model?

3. How is the concentration in units of molarity, i.e., the molar concentration, calculated in the model?

4. What other unit could you use for concentration rather than mol/L to specify the amount of substance in some volume of solution?

5. What is a situation that you have encountered where it was important to know a solution concentration?

6. Given the molar concentration of a solution, how can you determine the number of moles in some volume of that solution?

7. If a solution is diluted, say by adding an additional 100.0 mL of solvent, does the number of moles of solute change? Does the concentration increase, decrease, or stay the same?

8. Given the initial molar concentration and volume of a solution, how can you use the fact that the number of moles of solute present doesn't change when the solution is diluted to determine the molar concentration of the diluted solution?

EXERCISES

1. Calculate the molarity of a 0.175 L sugar solution that was prepared with 0.15 mole of sugar.

2. Determine the volume of 0.235 M sugar solution that can be prepared with 0.470 moles of sugar.

3. A salt solution is to be added to a marine aquarium. Calculate the molarity of a salt solution that is prepared by adding water to 18.65 g of NaCl to give a final volume of 250.0 mL.

4. The solution in Exercise 3 was diluted to 1.000 L. Calculate the molar concentration of the final solution.

PROBLEMS

1. Calculate the volume of a 3.15 M NaOH (aq) solution that should be used to prepare 250. mL of 0.150 M NaOH (aq).

2. Toxic chemicals in drinking water usually are reported and a safety level specified in units of parts per million (ppm) by mass. What is the molar concentration of arsenic in a water sample that has 1 ppm arsenic (As)?

Activity 05-3

Solving Solution Stoichiometry Problems

Why?

Stoichiometry, which comes from the Greek words for *element* and *measure*, is the process of tracking the elements and their masses in chemical reactions to ensure that reaction equations and calculations do not violate the conservation of mass. Scientists, engineers, and health professionals working with reactions in solution perform stoichiometric calculations to account quantitatively for the amounts of material that react and the amounts of material that are produced in the reactions. You will find these calculations straightforward if you use a strategy, such as the one in this activity that connects the reaction equation to the quantities of the chemical compounds in moles.

Learning Objectives

- Determine the amounts of reactants that react and products produced
- Learn the RICE table problem-solving strategy

Success Criterion

- Accurate calculations of the amount of material that reacts and is produced in a chemical reaction in solution

Prerequisites

- **Activity 03-3:** *Mole and Molar Mass*
- **Activity 04-1:** *Balanced Chemical Reaction Equations*
- **Activity 04-2:** *Dissociation and Precipitation Reactions*
- **Activity 04-3:** *Introduction to Acid – Base Reactions*
- **Activity 05-1:** *Limiting Reactants*
- **Activity 05-2:** *Solution Concentration and Dilution*

MODEL: RICE Table Problem-Solving Strategy

Calculate the mass of Ag_2CO_3(s) produced by mixing 125 mL of 0.315 M Na_2CO_3(aq) and 75.0 mL of 0.155 M $AgNO_3$(aq) and the number of moles of the excess reactant remaining in solution.

Your calculations will be more efficient if you realize that 1 M = 1mmol/mL and that molar masses can be expressed as mg/mmol as well as g/mol where 1mmol = 0.001 mol.

The RICE Table Solution

Steps in the Strategy	**Example** *(Note 1 mmol = 0.001 mol)*
Step R: Write the **Reaction equation**.	Both Na_2CO_3 and $AgNO_3$ dissociate in aqueous solution. The net ionic equation for the precipitation is: $2\,Ag^+(aq) + CO_3^{2-}(aq) \rightarrow Ag_2CO_3(s)$
Step I: Write the **Initial amounts** of reactants and products.	11.6 mmol 39.4 mmol 0 (75 mL)(0.155 M) = 11.6 mmol Ag^+ (125 mL)(0.315 M) = 39.4 mmol CO_3^{2-}
Step C: Write the **Change** in the amounts due to the reaction using the stoichiometric coefficients in the reaction equation.	In the reaction equation, the stoichiometric coefficient for silver is 2, the coefficient for carbonate is 1, and the coefficient for silver carbonate is 1. The equation is telling us that 2 moles of Ag^+ react with 1 mole of CO_3^{2-} to produce 1 mole of Ag_2CO_3. To get some amount of precipitate, say x moles, it will take x moles of carbonate and 2x moles of silver because the amounts must be in the same ratio as the coefficients (2:1:1 = 2x:x:x). $2\,Ag^+(aq) + CO_3^{2-}(aq) \rightarrow Ag_2CO_3(s)$ 2x x x
Step E: Write the amounts of substances present after the reaction reaches **Equilibrium** or is complete.	For the reactants, subtract the amounts that react from the initial amounts present. For the products, add the amounts produced to the initial amount present. $2\,Ag^+(aq) + CO_3^{2-}(aq) \rightarrow Ag_2CO_3(s)$ 11.6 − 2x 39.4 − x 0 + x
Then solve for unknowns as requested by the problem statement.	The final amount of the limiting reactant will be 0. The limiting reactant is silver so 11.6 − 2x = 0 and therefore x = 5.8 mmol. Now find how much carbonate remains and how much silver carbonate is produced. 39.4 – 5.8 = 33.6 mmol CO_3^{2-} remain. (5.8 mmol)(275.8 g/mol) = 1.60 g Ag_2CO_3 produced.

Key Questions

1. How are the amounts of reactants present initially calculated?

2. What does the abbreviation mmol represent?

3. Why is Ag^{+} and not CO_3^{2-} the limiting reactant?

4. Why is writing the balanced reaction equation an important part of the methodology?

5. Can you improve the methodology by changing the steps, changing the order of steps, or omitting some steps? Explain.

6. What insights about solving solution stoichiometry problems did your group gain from the Model and the Key Questions?

Exercises

1. Calculate the mass of $Fe(OH)_3(s)$ produced by mixing 50.0 mL of 0.153 M $KOH(aq)$ and 25.0 mL of 0.255 M $Fe(NO_3)_3(aq)$, and the number of moles of the excess reactant remaining in solution.

RICE Table Strategy Steps	Fill in the Details
R: Write the reaction equation.	$KOH + Fe(NO_3)_3 \rightarrow Fe(OH)_3$
I: Write the initial amounts of reactants and products.	
C: Write the change in amounts due to the reaction.	
E: Write the amounts after the reaction reaches equilibrium.	
Solve:	

2. Identify the excess reactant (HCl or NaOH), if any, and its final molar concentration when 1.55 g of NaOH(s) is stirred into 150.0 mL of 0.250 M HCl(aq).

RICE Table Strategy Steps	Fill in the Details
R: Write the reaction equation.	
I: Write the initial amounts of reactants and products.	
C: Write the change in amounts due to the reaction.	
E: Write the amounts after the reaction reaches equilibrium.	

Solve:

PROBLEM

1. One method of commercially removing the skins from potatoes is to soak them in a 3 to 6 M solution of sodium hydroxide for a short time. When the potatoes are removed from this solution, the skins are sprayed off. In one titration analysis of the NaOH solution, 36.2 mL of 0.650 M sulfuric acid (a diprotic acid) was required to neutralize (react completely) with a 25.0 mL sample of the soaking solution. Calculate the concentration of the NaOH soaking solution. Was it in the required range of 3 to 6 M?

RICE Table Strategy Steps	Fill in the Details
R: Write the reaction equation.	
I: Write the initial amounts of reactants and products.	
C: Write the change in amounts due to the reaction.	
E: Write the amounts after the reaction reaches equilibrium.	
Solve:	

Activity 06-1

Thermochemistry and Calorimetry

Why?

Chemical reactions release or store energy, usually in the form of *thermal energy*. Thermal energy is the kinetic energy of motion of the atoms and molecules comprising an object. Heat is thermal energy that is transferred from one object to another. The amount of heat released or stored by a chemical reaction can be determined from the change in temperature. You need to know how much heat is either released or taken up and stored by a reaction in order to determine the utility of that reaction. This also allows you to identify the conditions essential for that reaction to occur efficiently and safely.

Learning Objectives

- Quantify the relationship between heat and the change in temperature
- Understand the utility of the specific heat capacity
- Understand how to determine the heat of chemical reactions

Success Criteria

- Correctly use the specific heat capacity of substances in various situations to interrelate the amount of heat and the change in temperature
- Determine the heat of a chemical reaction from the change in temperature

Prerequisites

- **Activity 04-1:** *Balanced Chemical Reaction Equations*
- **Activity 04-3:** *Introduction to Acid – Base Reactions*

Information

Energy is measured in units of Joules (J).

Thermal energy is transferred spontaneously from an object at a higher temperature to another object at a lower temperature. Thermal equilibrium is reached when the two objects are at the same temperature.

The temperature of an object increases when it absorbs heat; the temperature of an object decreases when it releases heat. This temperature change can be used to determine the amount of heat that was absorbed or released if the relationship between these two quantities is known.

Model 1: Measuring Thermal Energy

In an experiment to determine the relationship between the temperature change of water and the amount of heat absorbed by the water, an electrical heater was used to increase the temperature of 1 g of water by 1 °C. The heat produced by an electrical heater over some period of time can be calculated very accurately: heat = electrical current × voltage × time. The results of this experiment are given below.

Mass of Water	Change in Temperature	Energy Used in Heating the Water
1.000 g	1.000 °C	4.184 J

Key Questions

1. How much heat was required to increase the temperature of 1.000 g of water by 1.000 °C?

2. How much heat would be required to increase the temperature of 100.0 g of water by 1.000 °C?

3. How much heat would be required to increase the temperature of 1.000 g of water by 50.0 °C?

4. The specific heat capacity of any substance is defined as the amount of energy needed to increase the temperature of 1 g of the substance by 1 °C. What is the specific heat capacity of water? Be sure to include the units of J/g °C.

5. Based on your answers to Key Questions 1-4, what is the equation that relates the amount of heat absorbed or released, by any substance, to the change in temperature of that substance? Write this equation using the following notation: q = heat, m = mass, ΔT = change in temperature, and c_s = specific heat capacity of the substance.

6. If you needed to experimentally determine the specific heat capacity of a new metal alloy that you developed, what would you need to measure, and how would you calculate the specific heat capacity from the measured values?

Exercises

1. Aluminum has a specific heat capacity of 0.902 J/g °C. How much energy is released when 1.0 kg of aluminum cools from 35 °C to 20 °C?

2. In preparing dinner you need a cup of very hot water, which you prepare on your electric stove. You use 80 kJ of electrical energy to heat 250 g of water starting at 20 °C. What is the final temperature of the water?

Got It!

1. At room temperature, equal masses of water and aluminum were each heated with 1 kJ of electrical energy. The aluminum became much hotter than the water. Explain why.

Model 2: Heat of a Chemical Reaction

Solutions of hydrochloric acid and sodium hydroxide are mixed in a beaker that is surrounded by Styrofoam insulation. The temperature of the solution and beaker increases from 24 °C to 38 °C. The volume of the resulting solution is 435 mL. Determine the amount of energy, ΔE_{total}, released by this reaction.

$$\Delta E_{total} = \Delta E_{solution} + \Delta E_{beaker}$$

ΔE_{total} = total energy released

$\Delta E_{solution}$ = energy used to heat the solution

$$\Delta E_{solution} = c_s \, m \, \Delta T$$

ΔE_{beaker} = energy used to heat the beaker, insulation, and thermometer

$$\Delta E_{beaker} = (330 \text{ J/°C}) \, \Delta T$$

Assume that since the solution is dilute, the specific heat of the solution is the same as that of water. The beaker, insulation, and thermometer have a combined heat capacity of 330 J/°C. Heat capacity differs from specific heat because heat capacity does not depend on the mass of anything. The *heat capacity* is the amount of energy required to change the temperature of something by 1°C.

$$\Delta E_{total} = (4.184 \text{ J/g °C})(435 \text{ mL} \times 1.00 \text{ g/mL})(14 \text{ °C}) + (330 \text{ J/°C})(14 \text{ °C}) = 30.1 \text{ kJ}$$

Key Questions

7. In **Model 2**, what things become hotter due to the energy that is released in the reaction of hydrochloric acid and sodium hydroxide?

8. How is the energy calculated that is used in heating the solution in **Model 2**?

9. How is the energy calculated that is used in heating the other items in **Model 2**?

10. How is the total energy released in the chemical reaction in **Model 2** calculated?

11. What is the difference between the specific heat capacity and the heat capacity of a substance or object?

PROBLEMS

1. You mix 100. mL of 1.0 M HCl with 100 mL of 1.0 NaOH, both at 25 °C. The temperature of your calorimeter rises by 5.98 °C, and its heat capacity is 100 J/°C. What is the heat of reaction per mol of $H_2O(l)$ formed?

2. An unknown piece of metal weighing 100. g is heated to 90 °C. It is dropped into 250 g of water at 20 °C. When equilibrium is reached, the temperature of both the water and piece of metal is 29 °C. Determine the specific heat of the metal using the fact that the heat lost by the metal must equal the heat absorbed by the water. Report all given temperature data as n.0 °C (i.e. 20.0 °C, 29.0 °C, 90.0 °C). Assume that the heat capacity of the container, a Styrofoam cup, is negligible.

Activity 06-2

Internal Energy and Enthalpy

Why?

The capability of a chemical system or machine to store and release energy or to do work is related to changes in the internal energy of the system or machine. The thermal energy stored or released when the pressure is kept constant is called the *enthalpy*. The concept of enthalpy will help you better understand the properties of chemical reactions and identify optimum systems for storing and releasing energy and producing work.

Learning Objectives

- Identify the relationships among internal energy, enthalpy, heat, and work
- Learn how to determine changes in internal energy and enthalpy through experimentation

Success Criteria

- Identify processes as endothermic or exothermic
- Determine the sign and magnitude of changes in internal energy and enthalpy

Prerequisite

- **Activity 06-1:** *Thermochemistry and Calorimetry*

Information

When physical and chemical changes occur in a system composed of atoms and molecules, energy is usually absorbed or released, and work is often done on or by the system. Energy is required to pull atoms apart and break chemical bonds. When bonds are formed, a more stable or lower energy situation is produced, so energy is released. When a chemical reaction involves both bond breaking and bond making, energy is either released or absorbed depending on the balance between the two. The energy change associated with a chemical reaction is usually in the form of thermal energy. The heat of a reaction that takes place at a constant pressure is referred to by chemists as the *enthalpy of reaction*. The important relationships among heat, work, enthalpy, and internal energy of a system are represented in the Model.

MODEL: WHAT AFFECTS THE INTERNAL ENERGY AND THE ENTHALPY?

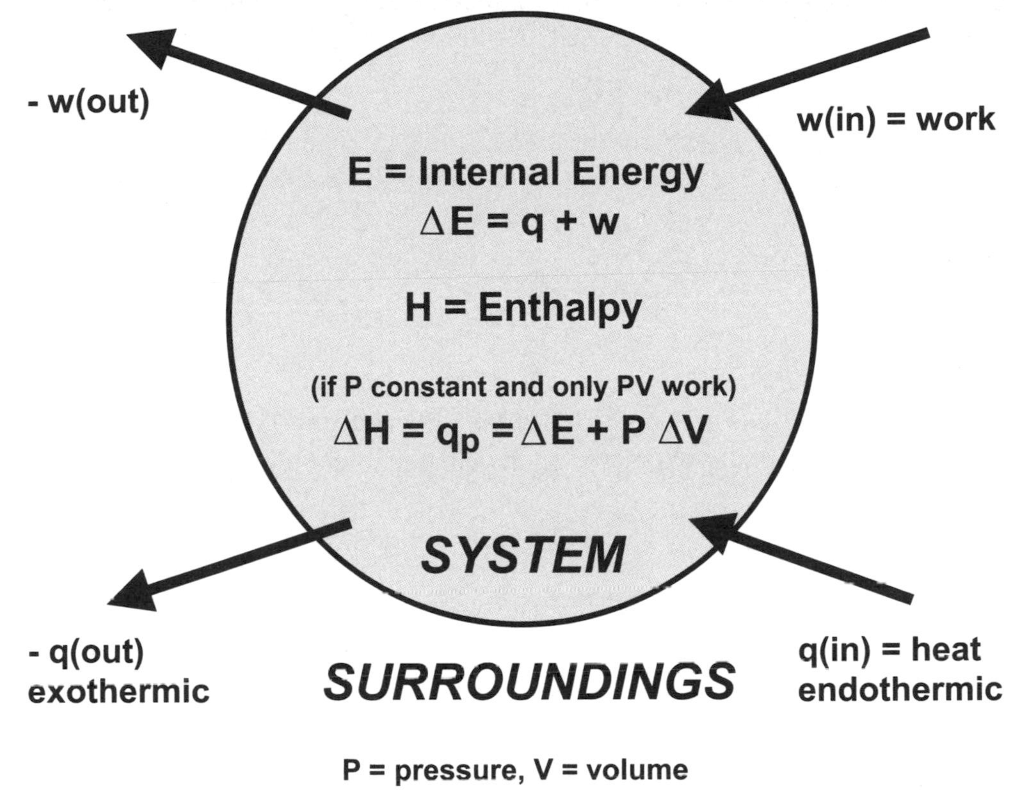

KEY QUESTIONS

1. According to the Model, if heat is introduced into a system from the surroundings, how does the internal energy of the system change?

2. According to the Model, if work is done on a system, e.g., by compressing it, how does the internal energy of the system change?

3. If a machine such as a steam engine produces work, how does the internal energy of the machine change?

4. Heat flows into or out of the system. What two terms describe the direction of heat flow? Write definitions of these terms.

5. What is the special name given to the heat (q_p) absorbed or released by a system when the pressure P is constant and the only work done is associated with a volume change?

6. In compressing a gas at constant pressure, work is done on the system and the internal energy increases. The work done is given by P times ΔV, where $\Delta V = V_{final} - V_{initial}$. Note that since the gas is compressed, V_{final} is smaller than $V_{initial}$. In calculating the change in internal energy, which of the following expressions should be used? Explain.

   a) $w = P\,\Delta V$ b) $w = -P\,\Delta V$

7. Since most chemical reactions are conducted in containers open to the atmosphere, why is the heat of a chemical reaction generally expressed as the change in enthalpy?

Exercises

1. If you do 1 J of work by pulling on a rubber band, by what amount does the internal energy of the rubber band change? Write your answer with a positive or negative sign as appropriate.

2. If your hot coffee loses 50 kJ of energy in cooling, what is the change in enthalpy of the coffee? Provide both the sign and the magnitude of ΔH.

3. Under what condition will the changes in enthalpy and internal energy be identical?

4. Identify each of the following as endothermic or exothermic. Explain.
   a) steam condensing

   b) ice melting

   c) two atoms combine to form a molecule: $2Cl(g) \longrightarrow Cl_2(g)$

   d) an electron is removed from an atom: $Na(g) \longrightarrow Na^+(g) + e^-$

e) a molecule is dissociated to produce two ions: $NaF(g) \longrightarrow Na^+(g) + F^-(g)$

5. Identify whether the sign of ΔH is positive or negative for each of the following. Explain. For any quantity, the change is determined by subtracting the initial from the final quantity, so

$$\Delta X = X_{final} - X_{initial}$$

a) steam condensing

b) ice melting

c) $Cl_2(g) \longrightarrow 2Cl(g)$

d) $Na(g) \longrightarrow Na^+(g) + e^-$

e) $Na^+(g) + F^-(g) \longrightarrow NaF(s)$

Problems

1. Your cup of hot coffee loses 50 kJ of energy in cooling, and the volume shrinks because of thermal contraction. For questions i) and ii), select an answer from the following list of possibilities, a) through f), and explain your reasoning.

a) exactly 50 kJ	c) less than 50 kJ	e) more negative than –50 kJ
b) more than 50 kJ	d) exactly –50 kJ	f) less negative than –50 kJ

i) How large is the change in enthalpy of the coffee?

ii) How large is the change in internal energy of the coffee?

iii) Justify your answers to i) and ii) by explaining why the change in internal energy must be larger/smaller than the change in enthalpy.

2. Burning butane (C_4H_{10}) produces gaseous carbon dioxide and water. The enthalpy of combustion of butane is – 2650 kJ/mole. Determine how much water you can heat from room temperature (22 °C) to boiling with 1 lb of butane.

3. You are at your cabin in the woods. Will you be able to take a hot bath tonight if only 0.1 lb of propane remains in the tank? Explain. (Use your textbook for any additional data or information that you may need, and you may find it necessary to make assumptions or approximations as well.)

ACTIVITY 06-3

Hess's Law: Enthalpy is a State Function

WHY?

The conditions that describe a system (such as temperature, pressure, volume, amount of material, and composition) are called its *state*. A quantity is called a *state function* when its value depends only on the state, and not on the path or route used to change it from one state to another. Since enthalpy is a state function, the change in enthalpy for a chemical reaction is the same whether the reaction takes place in one step or in a series of steps. This statement is known as Hess's Law. You can use this property of enthalpy to find the enthalpy change for a reaction from the enthalpy changes for other, related reactions. Hess's Law extends your ability to evaluate many chemical reactions for their potential to store and release energy and do work.

LEARNING OBJECTIVES

- Understand the concept of a state function
- Master the use of Hess's Law in calculations of reaction enthalpies

SUCCESS CRITERION

- Effectively use Hess's Law in calculations of reaction enthalpies

PREREQUISITE

- **Activity 06-2:** *Internal Energy and Enthalpy*

INFORMATION

Hess's Law means that the enthalpy change for any chemical reaction is equal to the sum of the enthalpy changes for any set of other reactions that lead to the same overall reaction. This idea is general and can be applied to any quantity that is a state function. A *state function* is defined as a property that depends only on the state of the system, which means that the property is independent of the history of the system or the path by which the system reached that state.

It is difficult to determine through experiment the enthalpy of formation of nitrogen monoxide, NO, because the usual product of nitrogen, N_2, reacting with oxygen, O_2, is nitrogen dioxide, NO_2, not NO. As illustrated in the following Model, Hess's Law makes it possible to combine the reaction leading to the formation of NO_2 (Reaction 1 in the Model) with the reaction converting NO to NO_2 (Reaction 2 in the Model) in order to obtain the enthalpy of formation of NO.

Model: Application of Hess's Law

Reaction 1: $N_2(g) + 2O_2(g) \longrightarrow 2NO_2(g)$ $\Delta H_1 = 68$ kJ

Reaction 2: $NO(g) + \frac{1}{2} O_2(g) \longrightarrow NO_2(g)$ $\Delta H_2 = -56$ kJ

Reaction 3: $\frac{1}{2} N_2(g) + \frac{1}{2} O_2(g) \longrightarrow NO(g)$ $\Delta H_3 = ?$ kJ

If you reverse Reaction 2, you obtain Reaction 4. If you multiply Reaction 1 by ½, you obtain Reaction 5. If you combine Reaction 4 with Reaction 5, you obtain Reaction 3 and a value for ΔH_3.

When you reverse a reaction, you change the sign of ΔH. When you multiply a reaction by constant you multiply ΔH by the same constant. And when you combine or add reactions, you add the ΔH values together.

Reaction 4: $NO_2(g) \longrightarrow NO(g) + \frac{1}{2} O_2$ $\Delta H_4 = 56$ kJ

Reaction 5: $\frac{1}{2} N_2(g) + O_2(g) \longrightarrow NO_2(g)$ $\Delta H_5 = 34$ kJ

Rxn 4 + Rxn 5: $NO_2(g) + \frac{1}{2} N_2(g) + O_2(g) \longrightarrow NO(g) + \frac{1}{2} O_2 + NO_2(g)$

$\frac{1}{2} N_2(g) + \frac{1}{2} O_2(g) \longrightarrow NO(g)$

$\Delta H_3 = \Delta H_4 + \Delta H_5 = -\Delta H_2 + \frac{1}{2} \Delta H_1 = 90$ kJ

Two paths for the formation of NO. The enthalpy change is independent of the path.

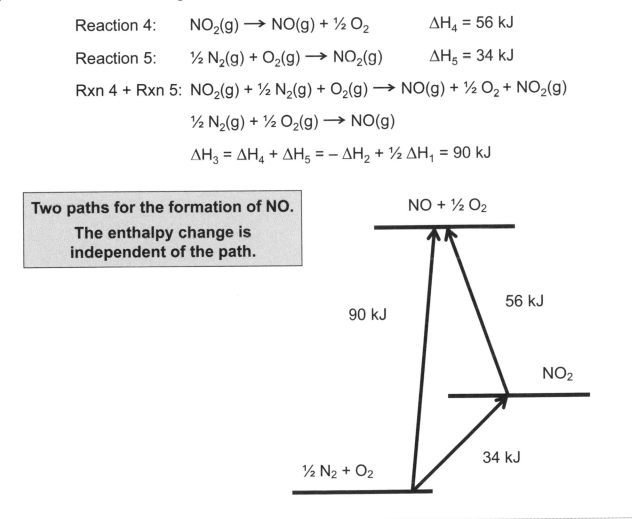

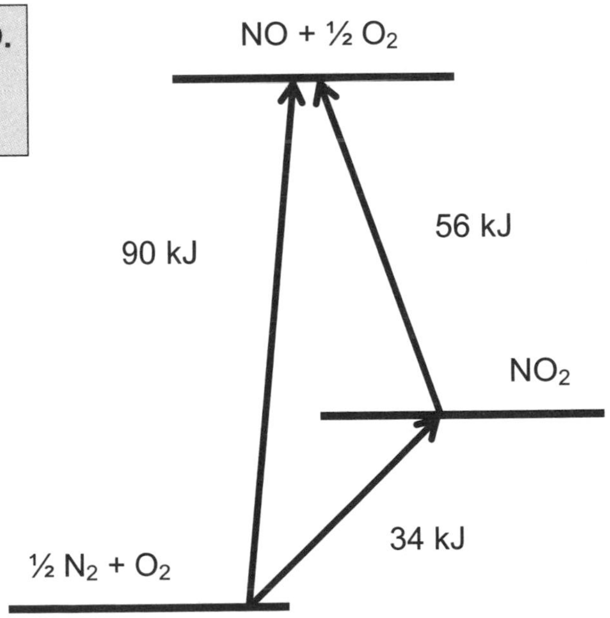

Key Questions

1. What is the enthalpy change for the reaction $NO + 1/2\ O_2 \longrightarrow NO_2$?

2. What is the enthalpy change for the reaction $NO_2 \longrightarrow NO + 1/2\ O_2$?

3. When a reaction is reversed, why is the absolute magnitude of the enthalpy change the same, but the sign of the enthalpy change different?

4. Why is the enthalpy change for Reaction 1 in the Model twice as large as the enthalpy change for Reaction 5?

5. How does the diagram in the Model illustrate that enthalpy is a state function?

6. Other than enthalpy, what is a quantity that you have encountered that is a state function? What is a quantity that is not a state function?

EXERCISE

1. Summarize how Reaction 3 and the associated enthalpy change, ΔH_3, are obtained from Reactions 1 and 2 in the Model. Explain why each step is necessary.

PROBLEMS

1. Calculate the standard enthalpy of formation (ΔH°_f) of ethanol, $C_2H_5OH(l)$, from the heat of combustion of ethanol, which is –1368 kJ/mole, by using tabulated standard enthalpies of formation for $CO_2(g)$ and $H_2O(l)$.

2. Draw a diagram, similar to the one in the Model, to illustrate that the enthalpies of formation of gaseous water and liquid water from gaseous dihydrogen and dioxygen are different, and indicate on the diagram what that difference is. Include the reaction equations and values for the enthalpy changes at 25 °C in your diagram. You can find enthalpy values in your textbook.

3. Use your diagram from Problem 2 to determine how much energy it takes to convert one mole of liquid water to the gas phase at 25 °C.

ACTIVITY 07-1

Electromagnetic Radiation

WHY?

Electromagnetic radiation, which is also called light, is an amazing phenomenon. It carries energy and has characteristics of both particles and waves. We can see only a small region of the electromagnetic spectrum, which we call *visible light*. The absorption and emission of electromagnetic radiation by atoms and molecules serve as powerful tools used to probe molecular structure and chemical reactions. They form the basis of medicine's magnetic resonance imaging and are intrinsic to many analytical techniques used to monitor the environment and manufacturing processes. Radio and television, cell phones, microwave ovens, and compact discs all utilize electromagnetic radiation.

LEARNING OBJECTIVE

- Characterize electromagnetic radiation

SUCCESS CRITERIA

- Interrelate the wavelength, frequency, momentum, and energy associated with electromagnetic radiation
- Identify the different regions of the electromagnetic spectrum

INFORMATION

During the nineteenth century, research in the areas of optics, electricity, and magnetism provided convincing evidence that electromagnetic radiation consists of two oscillating waves. One wave corresponds to an electric field, and the other wave corresponds to a magnetic field. In a vacuum, these fields oscillate perpendicular to each other and perpendicular to the direction the wave is moving. A wave is characterized by its amplitude, frequency, and wavelength. The model in this activity shows a diagram of an electromagnetic wave.

The Greek letter nu, ν, is used to represent frequency (cycles/s). Be careful to distinguish it from the English vee, **v**. Frequency is measured in hertz (Hz), which is expressed in cycles or oscillations per second.

The Greek letter lambda, λ, is used to represent wavelength.

During the twentieth century, scientists discovered that electromagnetic radiation also had properties normally associated with particles. This discovery led scientists to believe that electromagnetic radiation consists of particles called photons. A photon has a momentum, a specific amount of energy, and a wavelength and frequency associated with it. Thus, the properties of particles (momentum and a specific energy) and the properties of waves (wavelength and frequency) are blended together.

The wavelength and frequency of electromagnetic radiation extend essentially from 0 to infinity. The electromagnetic spectrum is viewed as split into different regions. These regions are determined by the nature of instrumentation (sources, wavelength selectors, and detectors) used in the different regions. The model in this activity also includes a chart of the electromagnetic spectrum.

Model: Properties of Electromagnetic Radiation

Figure 1 shows an *electromagnetic wave* with the magnetic field oscillating parallel to the z-axis, the electric field oscillating parallel to the y-axis, and the wave moving along the x-axis. The x, y, and z-axes are perpendicular to each other.

The *wavelength* is the distance between any two corresponding points, e.g., from one maximum of the electric field to the next.

The *frequency* is the number of wavelengths that pass a point on the x-axis each second.

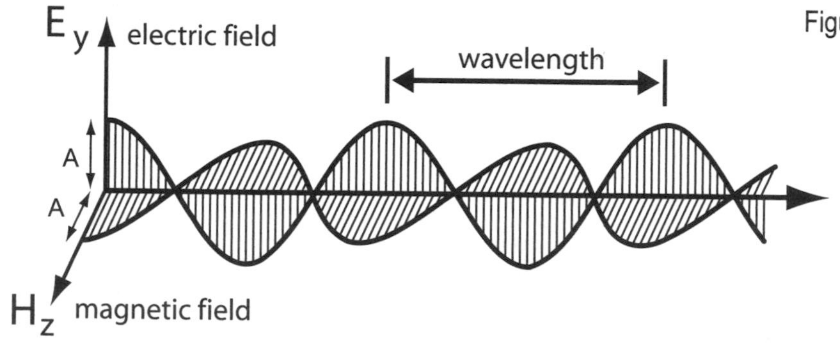

Figure 1

The chart shows the different spectral regions of the electromagnetic spectrum and indicates the approximate frequencies and wavelengths for each. The boundaries between the regions are diffuse.

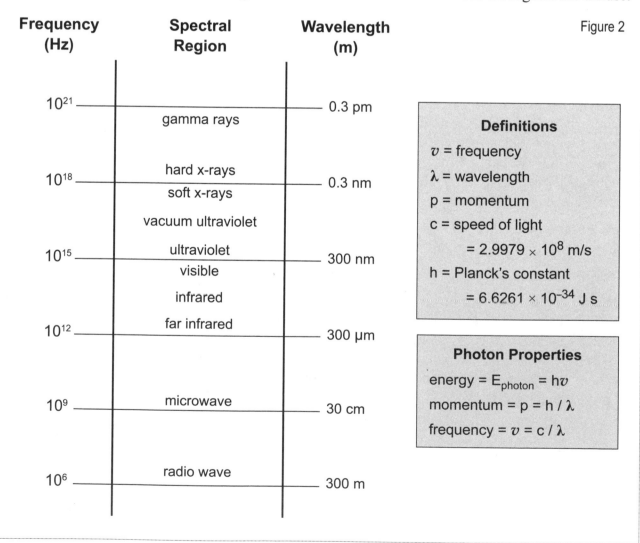

Figure 2

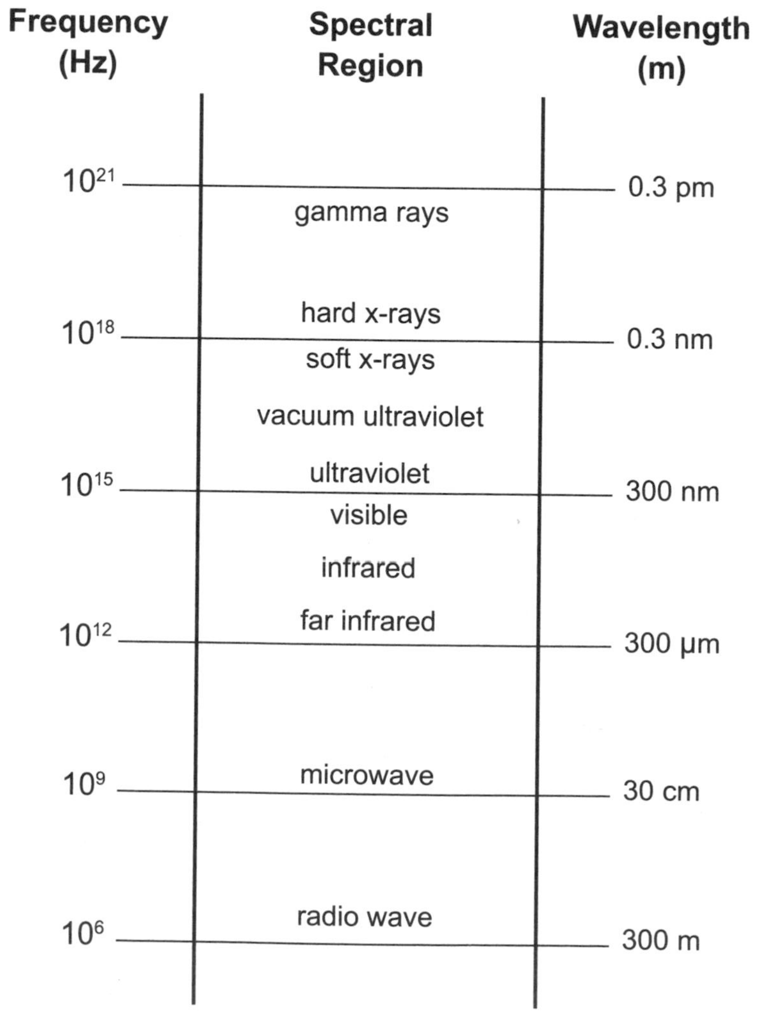

Definitions

v = frequency

λ = wavelength

p = momentum

c = speed of light

= 2.9979×10^{8} m/s

h = Planck's constant

= 6.6261×10^{-34} J s

Photon Properties

energy = $E_{photon} = hv$

momentum = $p = h / \lambda$

frequency = $v = c / \lambda$

Key Questions

1. In the Model, what is the equation showing the relationship between the energy (**E**) of a photon and the frequency of light (v)?

   E = hv

2. What is the equation showing the relationship between the frequency (v) and wavelength (λ) of light?

   λv = c

3. What is the equation showing the relationship between the momentum of a photon (**p**) and wavelength (λ)?

   h = λP

4. If two waves are traveling at the same speed along the x-axis, will the one with the longer wavelength have the larger or smaller frequency? Explain in terms of the number of wavelengths that pass a given point on the x-axis in 1 second.

   ~~larger~~ smaller

5. Is the energy of a photon proportional or inversely proportional to its frequency?

6. Is the momentum of a photon proportional or inversely proportional to its wavelength?

7. Which region of the electromagnetic spectrum has the shortest wavelengths?

8. Which region of the electromagnetic spectrum has photons with the lowest energy?

9. From what regions of the electromagnetic spectrum have you used or encountered photons? Identify the context.

Exercises

1. In the Model, draw a line connecting two points on the magnetic field wave that are separated by one wavelength.

2. The laser in a compact disc player uses light with a wavelength of 780 nm.

   a) Calculate the frequency of this light.

   3.83×10^{14} $\frac{2.99 \times 10^{8}}{780}$

   b) Calculate the energy of a single photon of this light.

   $(3.83 \times 10^{14})(6.626 \times 10^{-34}) = 2.54 \times 10^{-19}$

   c) Calculate the momentum of a single photon of this light.

   $p = \frac{h}{\lambda}$ $\frac{6.626 \times 10^{-34}}{780\left(\frac{1}{10^{9}}\right)}$

   8.495×10^{-28}

3. Radiation with a wavelength of 100 nm can be used to remove electrons from atoms and molecules. Identify the region of the spectrum corresponding to this radiation.

   Vacuum ultraviolet

4. Radiation emitted when excited states of nuclei decay has a frequency of approximately 10^{21} Hz. Identify the region of the spectrum corresponding to this radiation.

   gamma rays

5. Identify which uses photons with the higher energy: a microwave oven or a radio.

   microwave

Research

In different regions of the electromagnetic spectrum radiation is produced and detected in different ways and has different applications. If you are familiar with these different properties and characteristics, you will be able to assess safety issues, understand the limitations and opportunities in various applications, and even identify new applications. You can find information on spectroscopy on the internet and in library books on the subject.

Each team should prepare a report to the class on one region of the electromagnetic spectrum. This report should address the following items.

- An object about the size of one wavelength of this radiation
- A laboratory source of the radiation
- A method for obtaining tunable monochromatic radiation
- A device that can detect the radiation
- The effect on a molecule when it absorbs the radiation
- Documented effects of the radiation on the human body
- How the radiation is being used in modern research

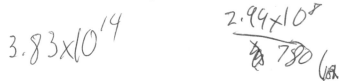

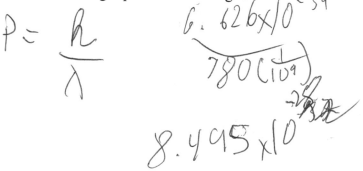

Activity 07-2

Atomic Spectroscopy and Energy Levels

Why?

The emission of light by the hydrogen atom and other atoms played a key role in helping scientists to understand the electronic structure of atoms. The light given off by atoms consists of narrow bands at specific wavelengths. The graph or other display of the light intensity as a function of wavelength is called an *emission* or *luminescence spectrum.* The spacing of the energies of the electrons in atoms can be obtained from luminescence spectra.

Learning Objective

- Understand how luminescence spectra can be related to the energy levels of electrons in atoms

Success Criteria

- Calculate the amount of energy absorbed or emitted by a hydrogen atom
- Relate luminescence bands to specific transitions between energy levels

Prerequisites

- **Activity 02-1:** *Atoms, Isotopes, and Ions*
- **Activity 07-1:** *Electromagnetic Radiation*

Information

A photon is produced when the electrons in an atom lose energy and make a transition from an upper energy level to a lower energy level.

Conservation of energy requires that the energy of the photon (**$h\nu$**) must equal the difference in energy between the two levels.

The Model shows a luminescence spectrum of the hydrogen atom in the vacuum ultraviolet region of the electromagnetic spectrum. The intensity of the light emitted (number of photons per second) is plotted on the y-axis, and the wavelength in nm is plotted on the x-axis.

This series of bands is called the *Lyman series*, after the physicist, Theodore Lyman, who first observed them. They are produced by transitions from excited states (higher energy levels) of the hydrogen atom to the ground state (the lowest energy level).

Each energy level is labeled with an index, $\boldsymbol{n}$, which is also called the *quantum number*. The quantum number $\boldsymbol{n}$ has integer values 1, 2, 3, etc. The energy of the hydrogen atom levels is related to the quantum number and is given by the following equation:

$E_n = -2.178 \times 10^{-18}\, Z^2 / \boldsymbol{n}^2$ Joules where Z is the atomic number for hydrogen, (Z=1).

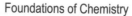

MODEL: LUMINESCENCE SPECTRUM AND ENERGY LEVELS OF THE HYDROGEN ATOM

Figure 1

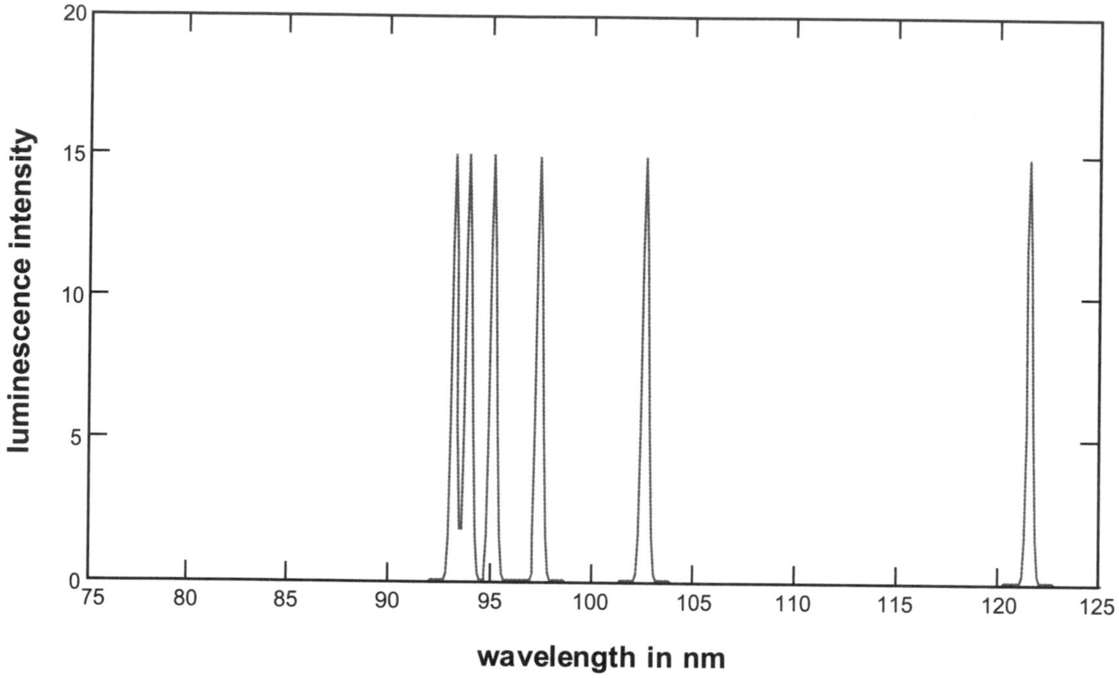

Figure 2

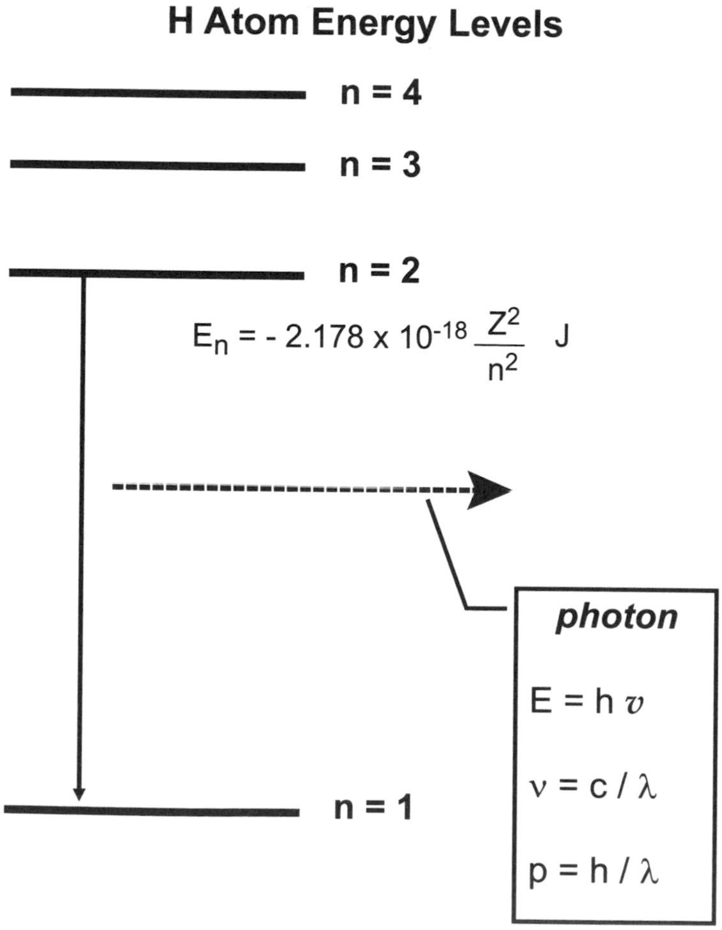

Key Questions

1. What is the equation that gives the possible energies (E_n) of the electron in the hydrogen atom?

2. What determines the energy of a photon that is emitted in the hydrogen atom luminescence?

3. How can the wavelength of a photon bc calculated from its energy?

Tasks

1. In the luminescence spectrum shown in the Model, assign the lines to transitions between the hydrogen atom energy levels. To aid in this assignment, complete Table 1 below. The first two rows have been completed for you. If you are working in a team of four, each person may do one calculation.

Table 1 Energies of H Atom Levels and Transitions

n (upper)	*n* (lower)	E(upper) in J	ΔE = E(upper) – E(lower)	λ in nm
2	1	-0.5445×10^{-18} J	1.634×10^{-18} J	121.6 nm
3	1	-0.2420×10^{-18} J	1.936×10^{-18} J	102.6 nm
4	1			
5	1			
6	1			
7	1			

2. Above each peak in the luminescence spectrum, write the quantum number of the upper level involved in the transition corresponding to that peak. Use your calculated wavelengths in Table 1 to help you make this assignment.

3. In the energy-level diagram below (which is based on Figure 2 from the Model), draw arrows to represent the transitions corresponding to the spectral bands. The arrow from ***n*** = 2 to ***n*** = 1 has been done for you. Label each arrow with the wavelength of the photon produced by that transition. Add additional energy levels to the diagram as needed, and label them with their values for the quantum number ***n***.

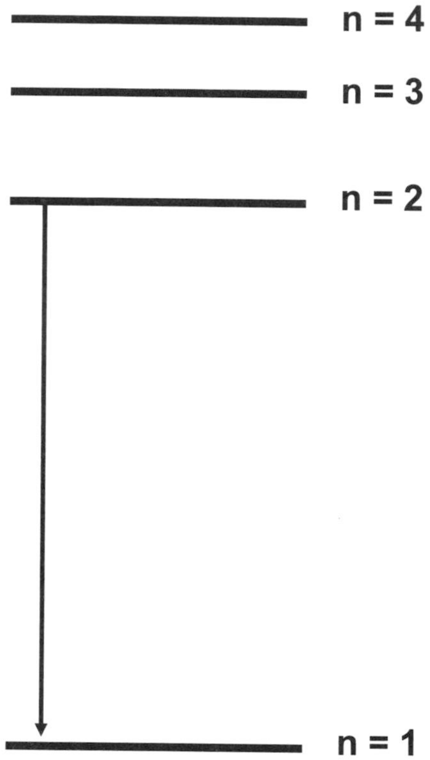

Got It!

1. Compare the transition that occurs when the electron in the hydrogen atom drops from energy level ***n*** = 2 to ***n*** = 1 to the transition that occurs from ***n*** = 3 to ***n*** = 1.

   a) Which transition causes the larger change in the energy of the hydrogen atom?

   b) Which transition will produce light with the longer wavelength?

2. Why doesn't the luminescence of the hydrogen atom produce light at all wavelengths?

Exercises

1. a) What is the wavelength of light that is absorbed when an electron in the hydrogen atom goes from the energy level $n = 2$ to $n = 4$?

$E_2 = -2.178 \times 10^{-18} \left(\frac{1}{2^2}\right) = -5.45 \times 10^{-19}$

$E_4 = -2.178 \times 10^{-18} \left(\frac{1}{4^2}\right) = -1.36 \times 10^{-19}$

$\frac{(6.626 \times 10^{-34})(2.99 \times 10^{8})}{4.09 \times 10^{-19}} = 4.84 \times 10^{-7}$ 4.09×10^{-19}

b) In which region of the spectrum is this light? hard x rays

c) Illustrate this transition in an energy level diagram similar to that used in the Model.

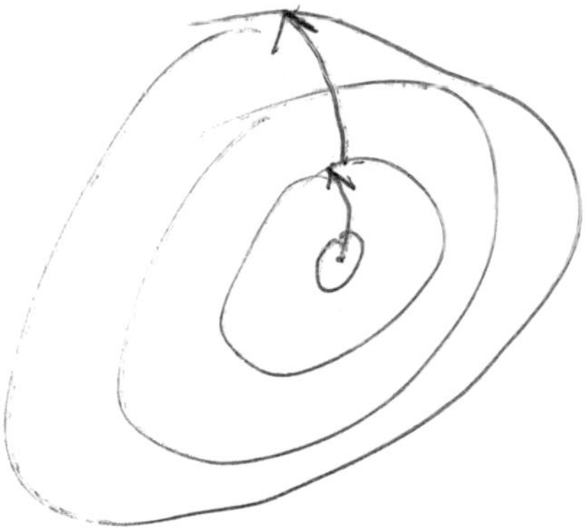

Problems

1. On Earth the ionization energy of atomic hydrogen is 1312 kJ/mol. On another planet the temperature is so high that essentially all the hydrogen atoms have the electron in the $n = 3$ quantum state. What is the ionization energy of atomic hydrogen on that planet?

2. Microwaves are used to heat food in microwave ovens. The microwave radiation is absorbed by moisture in the food. This absorption heats the water, and as it becomes hot so does the rest of the food. How many photons having a wavelength of 3.00 cm would have to be absorbed by 1.00 g of water to raise its temperature by 1.00 °C?

Activity 07-3

The Description of Electrons in Atoms

Why?

A description of electrons in atoms includes their energies and a quantitative picture of how they are distributed in space. This description is needed to understand how molecules are formed from atoms, how to synthesize new molecules, and how molecules function, i.e., the relationship between their structure, their reactivity, and their function in medicines, biological systems, and technological devices.

Learning Objective

- Gain an understanding of atomic orbitals

Success Criteria

- Correctly identify particular types of atomic orbitals
- Interrelate characteristics of atomic orbitals such as their name, shape, orientation in space, nodal pattern, extension in space, and energy

Prerequisite

- **Activity 07-2:** *Atomic Spectroscopy and Energy Levels*

Information

Realizing that light waves have momentum, and knowing that momentum is a property of particles, de Broglie proposed that particles, like electrons, must correspondingly have properties of waves. Experiments proved this to be true.

This success led Schrödinger to invent a mathematical equation that could be solved to provide wave functions describing electrons. This equation is now very famous and is called the Schrödinger equation.

While the Schrödinger equation is too complicated to be solved exactly for atoms with more than one electron, it can be solved exactly for atoms or ions with a single electron, like the hydrogen atom. The wave functions for the hydrogen atom are used to provide approximate wave functions for all other atoms. These approximate one-electron wave functions are called *atomic orbitals*.

While an atomic orbital is a mathematical function describing a single electron, chemists often think of the orbital as the region of space in which an electron can be found. An orbital can be represented by drawing a boundary surface to identify that the electron has a 90% probability of being within that surface. Such boundary surfaces for some atomic orbitals are shown in the Model.

While the energy levels of the hydrogen atom and ions with only one electron are determined by a single index or quantum number, n, Schrödinger discovered that more quantum numbers are involved in determining the energies, shapes, and orientations of the atomic orbitals. In addition to n, the quantum numbers ℓ and m_ℓ are also needed. These quantum numbers are called the *principal quantum number* (n), the *angular momentum* or *azimuthal quantum number* (ℓ), and the *magnetic quantum number* (m_ℓ).

MODEL: ATOMIC ORBITALS (WAVE FUNCTIONS) FOR ONE-ELECTRON ATOMS

Table 1

Quantum Numbers

Name	Characterizes	Symbol	Allowed Values
Principal	size energy total nodes = n–1	n	n = 1, 2, 3, … to infinity
Angular Momentum or Azimuthal	shape energy in multi-electron atoms planar nodes = ℓ	ℓ	ℓ = 0, 1, 2,…n –1
Magnetic	orientation	m_ℓ	$m_\ell = -\ell, -\ell+1, \ldots 0, \ldots \ell-1, \ell$ $2\ell + 1$ values

Figure 1

Energy Levels and Orbital Labels

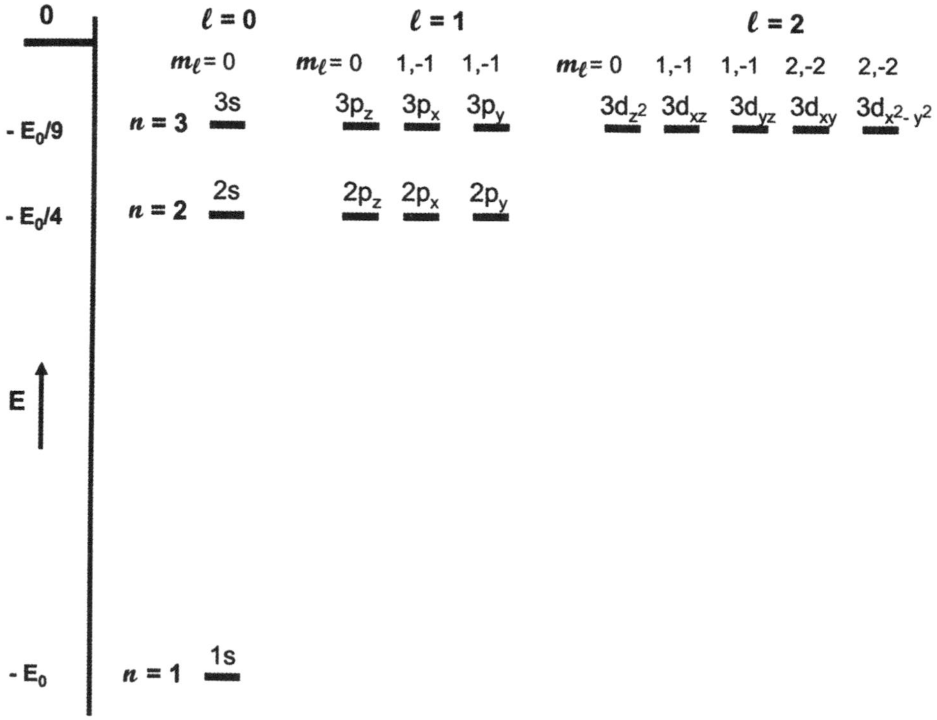

For atoms or ions with only 1 electron, orbital energies are ordered as follows:
1s < 2s=2p < 3s=3p=3d < 4s=4p=4d=4f < 5s=5p=5d=5f < 6s=6p=6d < 7s
For multi-electron atoms, orbital energies increase in the following order:
1s<2s<2p<3s<3p<4s<3d<4p<5s<4d<5p<6s<4f<5d<6p<7s<5f<6d

MODEL: ATOMIC ORBITALS (WAVE FUNCTIONS) FOR ONE-ELECTRON ATOMS (CONTINUED)

Figure 2

Shapes and Sizes of Atomic Orbitals

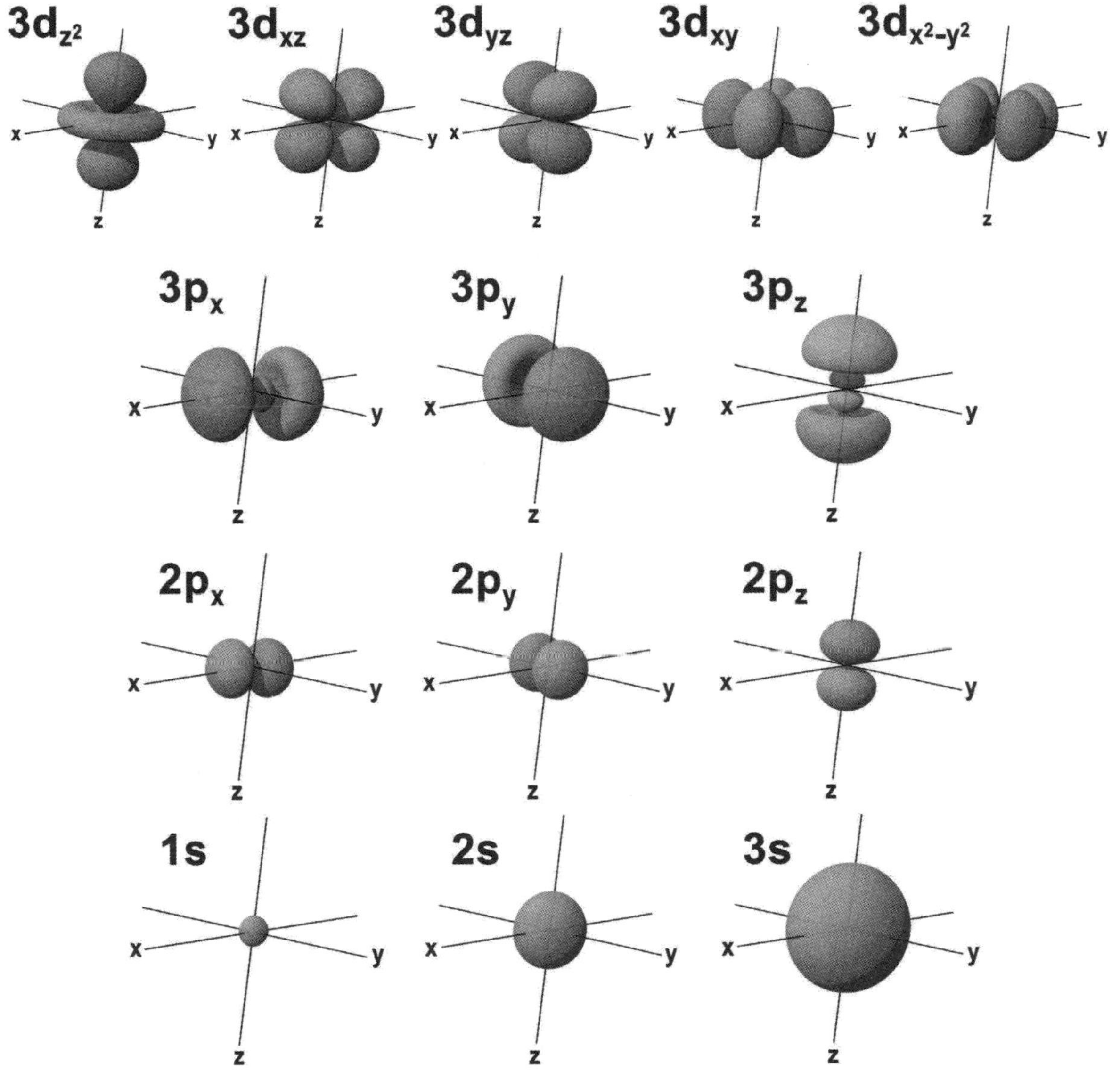

Figure 3

Radial Nodes in S-Orbitals

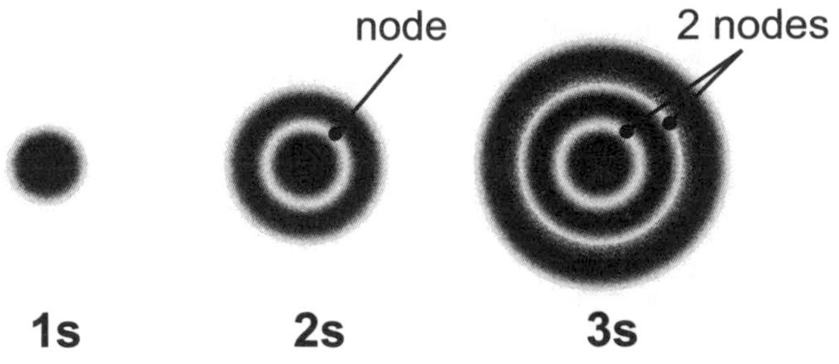

Key Questions

1. What are the characteristic shapes of s, p, and d orbitals that distinguish them from each other?

s=spherical
p=dumbol
d=cloverleaf

2. Which quantum number identifies the shape of an orbital? ℓ

3. What happens to the size of a particular type of orbital, as the principal quantum number increases? Consider s orbitals, for example.

it increases also

4. Which quantum numbers determine the energy of an orbital in an atom with more than one electron?

n and ℓ

5. For each value of ***n*** = 1, 2, and 3, what are the possible values for ***ℓ***, and what are the labels for the orbitals with these ***ℓ*** values? (For example s, p, or d.)

n	possible *ℓ* values	orbital labels
1	0	1s
2	1 0	2P 2s
3	2 1 0	3d 3P 3s

6. For each value of ***ℓ*** = 0, 1, and 2, what are the possible values for m_ℓ, and what are the labels, written as subscripts, for the orbitals with these m_ℓ values?

ℓ	possible m_ℓ values	orbital labels
0	0	s
1	-1 0 1	
2	-2 -1 0 1 2	

7. Which orbitals in the Model have a plane that includes the origin where the probability of finding the electron is zero? The *origin* is where the axes cross. These planes are called *angular nodal planes*. A *node* is the place where a wave has zero amplitude.

8. What is the relationship between the value of the azimuthal quantum number and the number of angular nodal planes?

9. In addition to angular nodal planes, there are *radial nodes*. These nodes occur for all angles at some distance from the nucleus. Which orbitals in the Model have radial nodes? Carefully compare the 2p and 3p orbitals before you answer this question.

Exercises

1. The total number of nodes in an orbital is equal to n–1. This number is split between radial nodes and angular nodal planes. Show that your answers to Key Questions 7 and 9 are consistent with this fact.

   n=1 = 0 nodes
   n=2 = 1 nodes
   n=3 = 2 nodes

2. For n = 4, identify the possible values for ℓ.

   3, 2, 1, 0

3. For ℓ = 3, identify the possible values for m_ℓ.

   -3, -2, -1, 0, 1, 2, 3

4. Identify the number of angular nodes in an ℓ = 3 orbital.

   3

5. Identify the following for the 1s, 2s, and 3s orbitals:

   n=1 n=2 n=3

   a) total number of nodes

   0, 1, 2

   b) number of angular nodes and the number of radial nodes

   1s = 0 2s = 1 3s = 2

   ℓ=0 ℓ=0 ℓ=0

6. Identify the relationship between the total number of nodes and the energy of the orbital.

   up

7. Identify the total number of nodes, the number of angular nodes, and the number of radial nodes for a 2p orbital.

$\ell = 1$ radial = 1
total = 2

8. Identify the total number of nodes, the number of angular nodes, and the number of radial nodes for a 3p orbital.

9. Identify the total number of nodes, the number of angular nodes, and the number of radial nodes for a 3d orbital.

10. Based on what you have learned so far about atomic orbitals, determine the total number of orbitals with $n = 4$

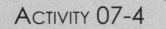

Activity 07-4

Multi-electron Atoms, the Aufbau Principle, and the Periodic Table

Why?

To construct the other elements from the simplest element, hydrogen, protons and neutrons are added to the nucleus and electrons are added to the orbitals in order of increasing energy. This is called the building-up principle (*aufbau* in German). By knowing the orbital energy-level structure and the number of electrons in an element, you can determine the electron configuration and thus the properties of the element. This information is summarized in the *Periodic Table of the Elements*, which you can use as a tool to identify materials with similar or contrasting chemical and physical properties.

Learning Objective

- Master the procedure for determining the electronic configuration of atoms and ions

Success Criteria

- Correctly write the electron configuration of atoms and ions
- Relate electron configuration to positioning in the Periodic Table

Prerequisites

- **Activity 07-2:** *Atomic Spectroscopy and Energy Levels*
- **Activity 07-3:** *The Description of Electrons in Atoms*

Model 1: Energy Level Diagram for the Hydrogen Atom

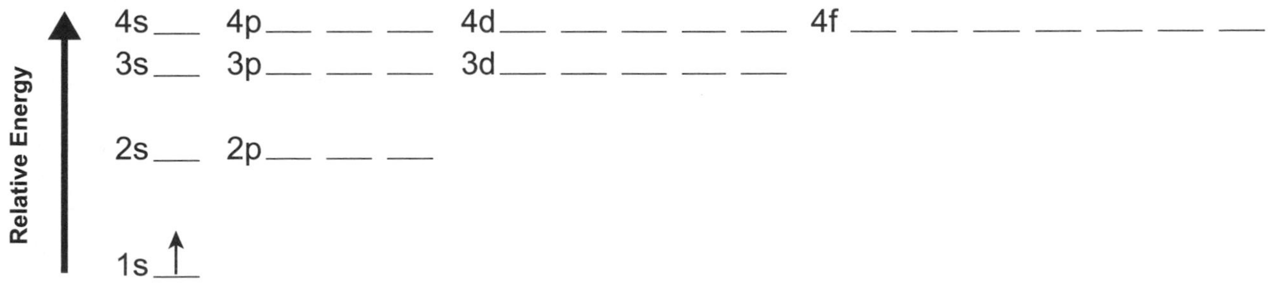

The electron configuration **1s**^{1} is a shorthand way to describe the electron in the hydrogen atom, which is represented by an arrow.

Suggestions contributed by Vicky Minderhout and colleagues at Seattle University

Key Questions

1. In writing the electron configuration for hydrogen as $1s^1$,

   a) What does the first 1 refer to?

   level

   b) What does the s refer to? shape

   c) What does the superscript 1 refer to?

   its hydrogen

2. Why are there different numbers of lines drawn after the symbols, e.g., 4s has one line, 4p has 3 lines, 4d has 5 lines and 4f has seven lines?

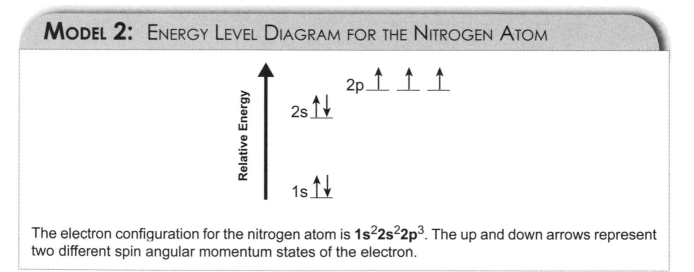

Model 2: Energy Level Diagram for the Nitrogen Atom

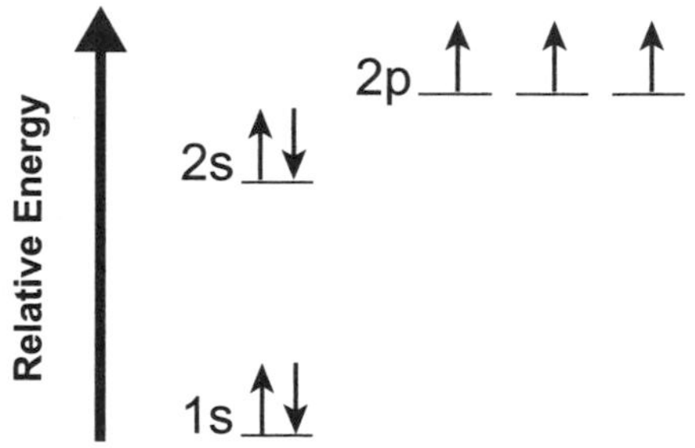

The electron configuration for the nitrogen atom is $\mathbf{1s^2 2s^2 2p^3}$. The up and down arrows represent two different spin angular momentum states of the electron.

Key Questions

3. What happens to the energies of the 2s and 2p orbitals when the atom has more than one electron?

4. In the notation $2p^3$, what information is provided by the 2, the p, and the 3?

5. What information about the electron is provided by the direction of the arrow in the diagram?

6. Are two electrons with the same spin angular momentum described by the same atomic orbital in nitrogen? Explain your answer.

7. If two atomic orbitals have the same energy, will the electrons pair up in one of them or go into different orbitals?

8. Explain why the three 2p electrons in nitrogen have the configuration $2p_x^1 2p_y^1 2p_z^1$ and not $2p_x^3$ or $2p_x^2 2p_y^1$.

Information

All atoms have the same kinds of orbitals as the hydrogen atom, but the energies of these orbitals change as the number of protons and number of electrons of the atom changes. The following guidelines are used to identify the orbitals that are occupied by electrons in any atom. These guidelines are derived from quantum mechanics and the solutions to Schrödinger's equation.

The *Aufbau Principle* says that, as protons are added to the nucleus to build up the elements, electrons are added first to the lowest energy atomic orbitals available before they fill higher energy orbitals. The name derives from *aufbau* in German, which means *building-up*. The Aufbau Principle makes sense because it produces the lowest energy or most stable arrangement of the electrons in the atom.

The *Pauli Exclusion Principle* says that two electrons cannot have the same set of quantum numbers simultaneously. This principle determines the number of electrons that can occupy each orbital. In addition to the three quantum numbers introduced in *Activity 07-3*: *The Description of Electrons in Atoms*, there is a fourth quantum number that is attributed to electron angular momentum. Electron spin angular momentum is a property that is described by Dirac's relativistic formulation of quantum mechanics. In some ways it is similar to a basketball or tennis ball spinning, yet there are important differences. Electron spin angular momentum is best regarded as the intrinsic angular momentum carried by an electron and not as the electron actually spinning. Dirac showed that each electron can have two values for the spin angular momentum quantum number, m_S: +½ and –½. So two electrons can be in each orbital and have the same values for n, ℓ, and m_ℓ, because they have different values for m_S. The two different spin angular momentum states for electrons are represented in orbital diagrams by up and down arrows or by the values +½ and –½ for the spin quantum number, m_S.

Hund's Rule says that if multiple orbitals with the same energy are available, then the unoccupied orbitals will be filled by electrons with the same spin before electrons with different spins pair up in occupied orbitals. Hund's Rule makes sense because electrons repel each other. If they are in the same orbital, they will be close together and their energy will be higher than it would be if they were separated in different orbitals.

EXERCISES

1. Complete the energy level diagram below and write the electron configuration for sulfur.

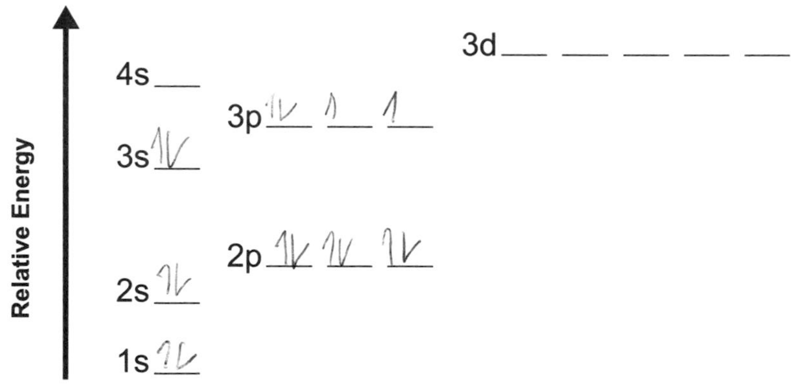

2. In your energy level diagram for sulfur, above, identify features that illustrate
   a) the Aufbau Principle

   b) the Pauli Exclusion Principle

c) Hund's Rule.

3. Complete the energy level diagram below and write the electron configuration for iron.

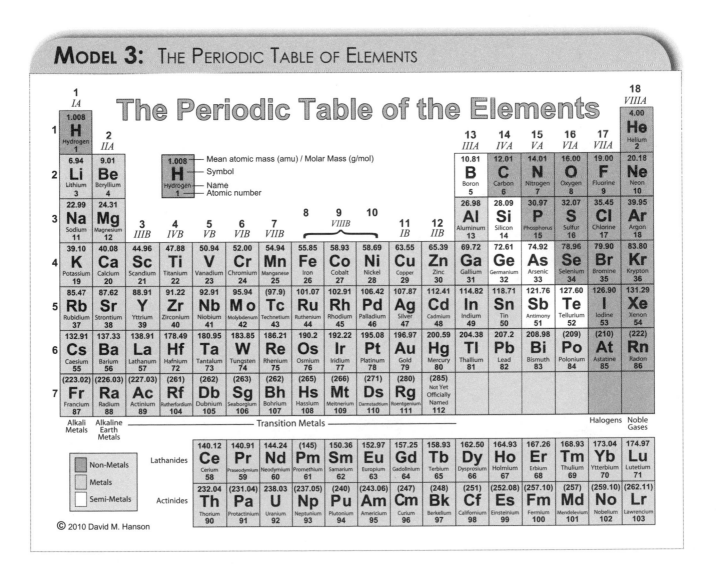

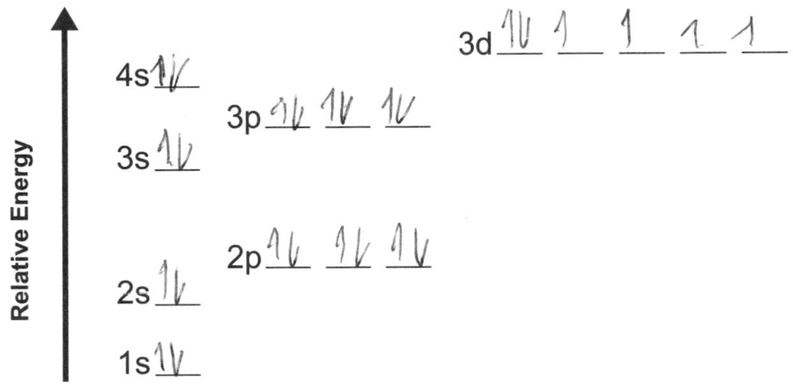

MODEL 3: THE PERIODIC TABLE OF ELEMENTS

The Periodic Table of the Elements

Key: 1.008 / H / Hydrogen / 1 — Mean atomic mass (amu) / Molar Mass (g/mol); Symbol; Name; Atomic number

	1 IA	2 IIA	3 IIIB	4 IVB	5 VB	6 VIB	7 VIIB	8	9 VIIIB	10	11 IB	12 IIB	13 IIIA	14 IVA	15 VA	16 VIA	17 VIIA	18 VIIIA
1	1.008 H Hydrogen 1																	4.00 He Helium 2
2	6.94 Li Lithium 3	9.01 Be Beryllium 4											10.81 B Boron 5	12.01 C Carbon 6	14.01 N Nitrogen 7	16.00 O Oxygen 8	19.00 F Fluorine 9	20.18 Ne Neon 10
3	22.99 Na Sodium 11	24.31 Mg Magnesium 12											26.98 Al Aluminum 13	28.09 Si Silicon 14	30.97 P Phosphorus 15	32.07 S Sulfur 16	35.45 Cl Chlorine 17	39.95 Ar Argon 18
4	39.10 K Potassium 19	40.08 Ca Calcium 20	44.96 Sc Scandium 21	47.88 Ti Titanium 22	50.94 V Vanadium 23	52.00 Cr Chromium 24	54.94 Mn Manganese 25	55.85 Fe Iron 26	58.93 Co Cobalt 27	58.69 Ni Nickel 28	63.55 Cu Copper 29	65.39 Zn Zinc 30	69.72 Ga Gallium 31	72.61 Ge Germanium 32	74.92 As Arsenic 33	78.96 Se Selenium 34	79.90 Br Bromine 35	83.80 Kr Krypton 36
5	85.47 Rb Rubidium 37	87.62 Sr Strontium 38	88.91 Y Yttrium 39	91.22 Zr Zirconium 40	92.91 Nb Niobium 41	95.94 Mo Molybdenum 42	(97.9) Tc Technetium 43	101.07 Ru Ruthenium 44	102.91 Rh Rhodium 45	106.42 Pd Palladium 46	107.87 Ag Silver 47	112.41 Cd Cadmium 48	114.82 In Indium 49	118.71 Sn Tin 50	121.76 Sb Antimony 51	127.60 Te Tellurium 52	126.90 I Iodine 53	131.29 Xe Xenon 54
6	132.91 Cs Caesium 55	137.33 Ba Barium 56	138.91 La Lathanum 57	178.49 Hf Hafnium 72	180.95 Ta Tantalum 73	183.85 W Tungsten 74	186.21 Re Rhenium 75	190.2 Os Osmium 76	192.22 Ir Iridium 77	195.08 Pt Platinum 78	196.97 Au Gold 79	200.59 Hg Mercury 80	204.38 Tl Thallium 81	207.2 Pb Lead 82	208.98 Bi Bismuth 83	(209) Po Polonium 84	(210) At Astatine 85	(222) Rn Radon 86
7	(223.02) Fr Francium 87	(226.03) Ra Radium 88	(227.03) Ac Actinium 89	(261) Rf Rutherfordium 104	(262) Db Dubnium 105	(263) Sg Seaborgium 106	(262) Bh Bohrium 107	(265) Hs Hassium 108	(266) Mt Meitnerium 109	(271) Ds Darmstadtium 110	(280) Rg Roentgenium 111	(285) Not Yet Officially Named 112						

Alkali Metals | Alkaline Earth Metals | Transition Metals | Halogens | Noble Gases

Lathanides	140.12 Ce Cerium 58	140.91 Pr Praseodymium 59	144.24 Nd Neodymium 60	(145) Pm Promethium 61	150.36 Sm Samarium 62	152.97 Eu Europium 63	157.25 Gd Gadolinium 64	158.93 Tb Terbium 65	162.50 Dy Dysprosium 66	164.93 Ho Holmium 67	167.26 Er Erbium 68	168.93 Tm Thulium 69	173.04 Yb Ytterbium 70	174.97 Lu Lutetium 71
Actinides	232.04 Th Thorium 90	(231.04) Pa Protactinium 91	238.03 U Uranium 92	(237.05) Np Neptunium 93	(240) Pu Plutonium 94	(243.06) Am Americium 95	(247) Cm Curium 96	(248) Bk Berkelium 97	(251) Cf Californium 98	(252.08) Es Einsteinium 99	(257.10) Fm Fermium 100	(257) Md Mendelevium 101	(259.10) No Nobelium 102	(262.11) Lr Lawrencium 103

Legend: Non-Metals; Metals; Semi-Metals

Model 3: The Periodic Table of Elements (continued)

Electron Configurations

Table 1

Element	Number of Electrons	Electron Configuration
He	2	$1s^2$
Li	3	$1s^2 2s^1$
Be	4	$1s^2 2s^2$
B	5	$1s^2 2s^2 2p^1$
Ne	10	$1s^2 2s^2 2p^6$
Na	11	$[Ne]3s^1$
Al	13	$[Ne]3s^2 3p^1$
Ar	18	$[Ne]3s^2 3p^6$

Element	Number of Electrons	Electron Configuration
K	19	$[Ar]4s^1$
Sc	21	$[Ar]4s^2 3d^1$
Ti	22	$[Ar]4s^2 3d^2$
Zn	30	$[Ar]4s^2 3d^{10}$
Ga	31	$[Ar]4s^2 3d^{10} 4p^1$
Kr	36	$[Ar]4s^2 3d^{10} 4p^6$
Rb	37	$[Kr]5s^1$

Key Questions

The rows in the Periodic Table are called periods, the columns are called groups, and the elements can be grouped together in blocks. All of this structure is due to and reflects the electron configurations of the atoms. The information in the Electron Configuration Table in **Model 3** (Table 1) and the following questions should help you deepen your understanding of the Periodic Table.

9. In terms of electron configurations, why are there only two elements in Period 1?

10. In terms of electron configurations, why do Periods 2 and 3 each contain eight elements?

11. In terms of electron configurations, why are there 18 elements in Period 4?

12. A block is formed by all the elements in Groups 1 and 2. What do these elements have in common?

13. A block is formed by all the elements in Groups 3 through 12. What do these elements have in common?

14. A block is formed by all the elements in Groups 13 through 18. What do these elements have in common?

15. A block is formed by the 28 elements set off at the bottom of the table. What do these elements have in common?

16. Why are the chemical properties of Na, K, and Rb similar?

17. In terms of their electron configurations, why do you think the elements in Group 18 are very stable, inert, and unreactive?

18. What do the elements in each group have in common?

Exercises

4. Identify three elements not in Groups 1 or 18 that you expect to have similar chemical properties. Explain why.

Be Ca Mg

because they are s orbitals

5. Identify the element with the following electron configuration: [Xe]$6s^2 4f^{14} 5d^{10}$.

Hg ~~Cd~~

6. Write the electron configuration for each of the following atoms or ions.

O	$1s^2 2s^2 2p^4$
Mg	
Mg^{2+}	
Cl	
Cl^-	
N^{3-}	

Activity 07-5

Periodic Trends in Atomic Properties

Why?

Many properties of atoms have a repeating pattern when plotted with respect to atomic number. The similarities are due to the repeating pattern of electron configurations involving s, p, d, and f orbitals. These experimentally observed periodic trends actually provide evidence for the orbital and shell structure of atoms as well as the meaningful arrangement of elements in the Periodic Table. Your ability to recognize the properties of the elements from their positions in the Periodic Table will prove useful in handling chemical compounds safely, developing new materials, finding new applications of known materials, or using chemistry in medical applications.

Learning Objective

- Develop relationships between position in the Periodic Table, electron configurations, atomic radius, ionization energy, and electron affinity

Success Criteria

- Use electron configurations and position in the Periodic Table to account for the relative sizes, ionization energies, and electron affinities of different atoms
- Order elements by atomic radius, ionization energy, and electron affinity

Prerequisites

- **Activity 07-3:** *The Description of Electrons in Atoms*
- **Activity 07-4:** *Multi-electron Atoms, the Aufbau Principle, and the Periodic Table*

Information

Properties such as the size of an atom (atomic radius), the energy required to remove an electron from an atom (ionization energy), and the energy required to remove an electron from a negative atomic ion (electron affinity) can be understood in terms of the electron configuration of the atom and the balance between *electron-nucleus attraction* and *electron-electron repulsion*. An electron in an atom is attracted by the positively charged nucleus and repelled by the other electrons. How they balance depends on how effective the electrons are in getting close to each other or close to the nucleus. If the electron-nucleus attraction (e-n) has a large effect, then the atom is small, has a high ionization energy, and large electron affinity. If the electron-electron repulsion (e-e) has a large effect, then the atom is large, the ionization energy is small, and the electron affinity is very small or zero. Table 1 summarizes what happens in the progression from one atom to the next in the Periodic Table; both the nuclear charge and the number of electrons increase by 1 unit.

Table 1 Effects of Attractive and Repulsive Interactions

Interaction	Atomic Size	Ionization Energy	Electron Affinity
e-n attraction > e-e repulsion	Decreases	Increases	Increases
e-n attraction < e-e repulsion	Increases	Decreases	Decreases

Model 1: Review of Electron Configurations of Atoms

The electron configuration of an atom specifies the number of electrons in each atomic orbital.

Example: The Electron Configuration of Carbon

$$1s^2 2s^2 2p^2$$

This notation for the electron configuration means that 2 electrons are described by or are in the 1s orbital, 2 are in the 2s orbital, and 2 are in 2p orbitals. Electron configurations can also be represented by orbital box diagrams as shown below. In these diagrams an electron is represented by an upward or a downward pointing arrow. An upward pointing arrow indicates that the electron has positive spin angular momentum in one direction (referred to as *spin up*). A downward pointing arrow indicates that it has negative spin angular momentum in that direction (referred to as *spin down*). Dirac showed that the idea of spin angular momentum is a consequence of Einstein's theory of special relativity and does not mean that the electron is spinning like a top.

The following Key Questions will help you identify the principles or rules that account for the electron configurations by helping you examine the orbital box diagrams in Figure 1.

Figure1

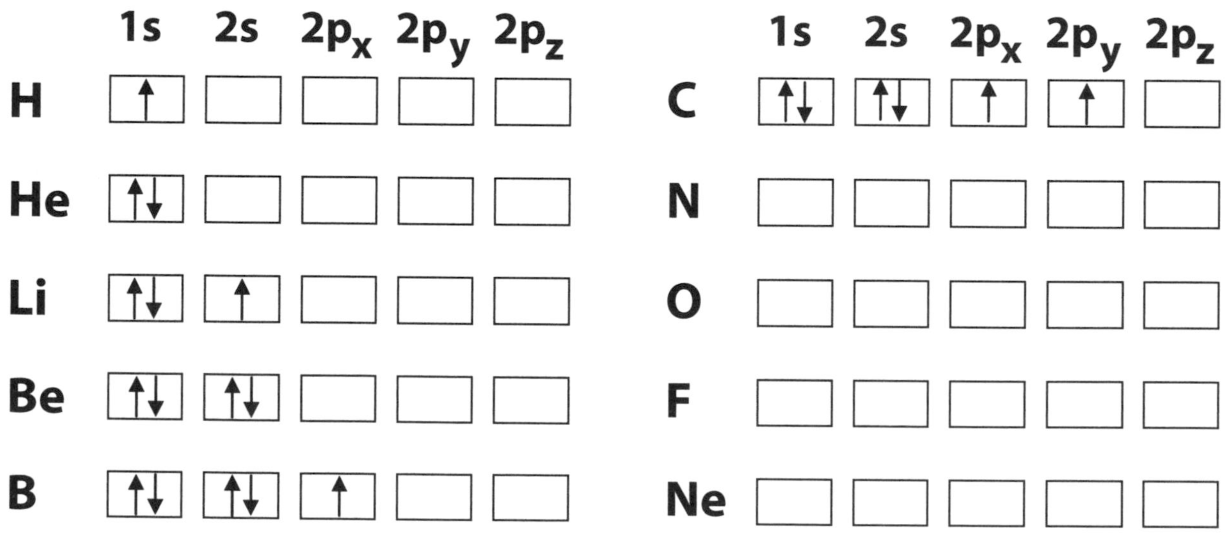

Key Questions

1. Consider the orbital box diagrams in **Model 1**. Do electrons fill the lower or higher energy orbitals first?

   lower

2. What is the maximum number of electrons that can go in each orbital?

2

3. If multiple orbitals with the same energy are available, e.g., in the case of 2p-orbitals, do the electrons all go in one orbital or do they go in different orbitals?

different orbitals

4. According to the information in **Model 1**, how many different spin angular momentum states are there for each electron?

2

KEY QUESTIONS

5. What are two ways shown in **Model 1** to write or specify the electron configuration of an atom?

6. What are the names of the three guidelines that determine the electron configuration of an atom?

7. In the orbital box diagram for carbon shown in **Model 1**, why aren't both p-electrons in the p_x-orbital?

Pa

8. Why isn't the electron configuration of carbon $1s^32s^3$?

$1s^2 2s^2 2p^2$

Exercises

Complete the orbital box diagrams in **Model 1** by applying the Aufbau Principle, the Pauli Exclusion Principle, and Hund's Rule.

Model 2: Electron-Electron and Electron-Nucleus Interactions in Atoms

Figure 2

Dominant Interparticle Forces in Helium

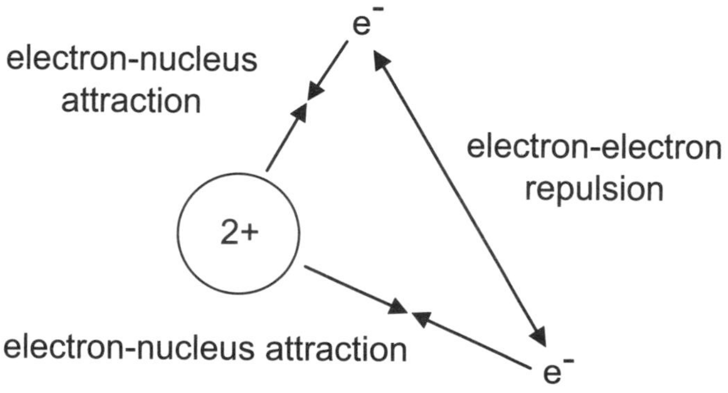

Figure 3

Dominant Interparticle Forces in Lithium

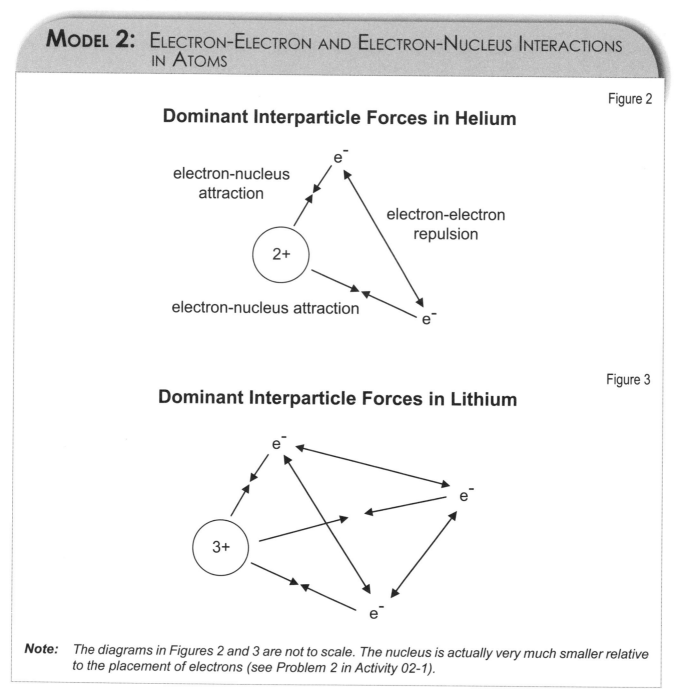

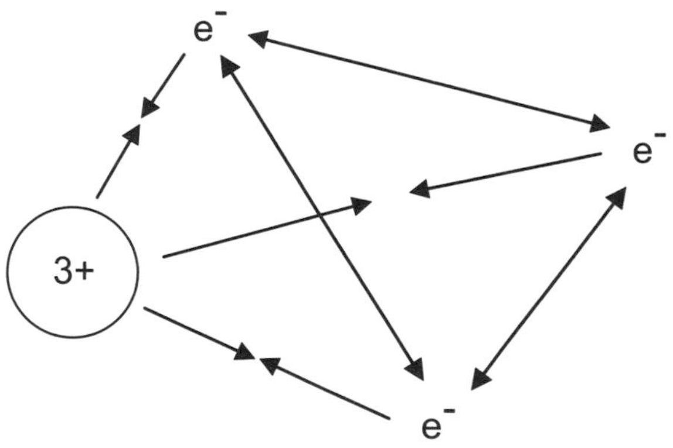

Note: *The diagrams in Figures 2 and 3 are not to scale. The nucleus is actually very much smaller relative to the placement of electrons (see Problem 2 in Activity 02-1).*

Information

The diagrams in **Model 2** illustrate how electron-electron repulsion can effectively reduce the attraction of an electron to the nucleus, and how this shielding effect depends on the electron configuration.

In helium, both of the electrons are in the same atomic orbital (1s) and therefore do not shield each other very well from the +2 charge of the nucleus. Consequently, the two electrons are strongly attracted to the nucleus.

In lithium, the outer electron is in a 2s orbital and is shielded very effectively from the +3 charge of the nucleus by the two electrons in the inner 1s orbital. As a result of this shielding and because it is farther from the nucleus, the outer electron experiences a smaller nuclear charge. Consequently, in lithium, the outer electron is not strongly attracted to the nucleus.

Key Questions

9. Is the electron-nucleus interaction attractive or repulsive? Explain.

10. Is the electron-electron interaction attractive or repulsive? Explain.

11. Based on your interpretation of Figures 2 and 3, do you agree or disagree with the following statements? Explain your reasons for agreeing or disagreeing.

    a) Electrons in the same shell do not shield each other from the nuclear charge very effectively. So, for the case of helium, their attraction to the nucleus is characteristic of a nuclear charge less than, but close to, +2.

    b) Electrons in inner shells are very good at shielding outer electrons from the nuclear charge. So, for the case of lithium, the electron-electron repulsion reduces the effective nuclear charge to a value much less than +3.

Information

Ionization energy is defined as the energy required to ionize an atom and corresponds to the following reaction equation. (Ionization energy is a positive quantity.)

$$X \longrightarrow X^{+} + e^{-}$$

By analogy, research scientists define *electron affinity* as the energy required to ionize an atomic anion. It corresponds to the following reaction equation. (Electron affinity is a positive quantity.)

$$Y^{-} \longrightarrow Y + e^{-}$$

In some general chemistry textbooks, the electron affinity is defined as the energy associated with the following electron attachment reaction.

$$Y + e^{-} \longrightarrow Y^{-}$$

Because this formula is just the reverse of the anion ionization reaction, this textbook convention produces an electron affinity with the same magnitude but a negative sign compared to the one used by research scientists.

Key Questions

12. Will an increase in the attraction to the nucleus tend to increase or decrease each of the following? Explain.
    a) The size of an atom ~~increase~~ decrease

    b) The ionization energy decrease

    c) The electron affinity increase

13. Will an increase in the repulsion of an electron by other electrons tend to increase or decrease each of the following? Explain.
    a) The size of an atom increase

    b) The ionization energy increase

    c) The electron affinity decrease

Exercises

2. In going from hydrogen to helium, there is a change in the atomic radius (from 37 to 32 pm) and a change in ionization energy (from 1311 kJ/mole to 2377 kJ/mole). Identify what these changes suggest about the relative magnitudes of the changes in the *electron–nucleus attraction* and the *electron–electron repulsion*.

3. Using the ideas illustrated in **Model 2** and developed by the preceding Key Questions and Exercise 2, explain why you would you expect lithium to be larger and have a smaller ionization energy than helium.

Model 3: Variation of Atomic Properties with Atomic Number

The unit for atomic radii is pm; the unit for ionization energies and electron affinities is kJ/mole.

Figure 4

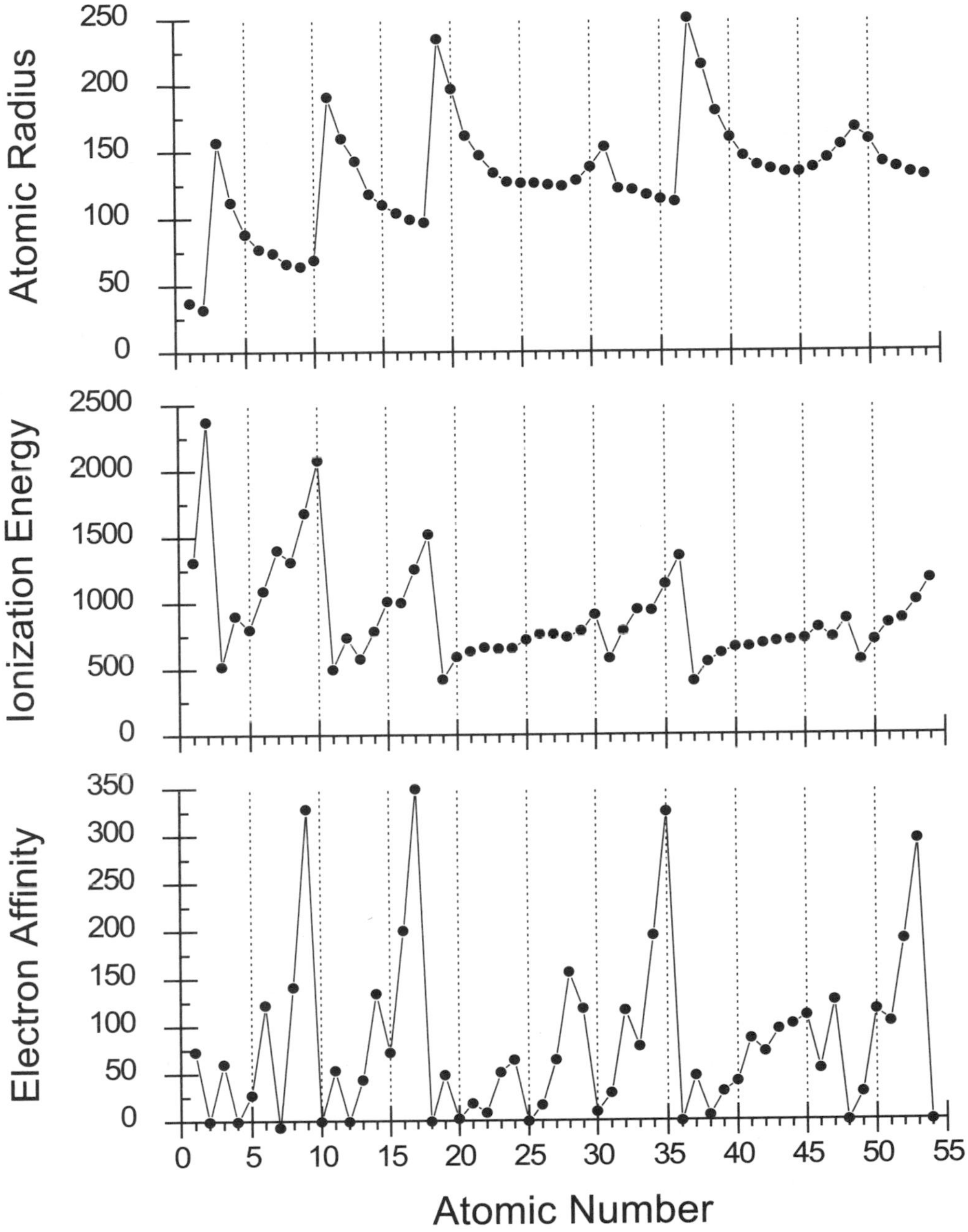

Key Questions

14. The Periodic Table is organized with the atomic number of the elements increasing from the beginning to the end. Similarly, the x-axis for all graphs in **Model 3** is the atomic number of the element. In going from one element to the next across the x-axis, what happens to the number of protons and the number of electrons?

protons increase
electrons increase

15. Do additional protons tend to increase or decrease the electron-nucleus attraction? Explain.

16. Do additional electrons tend to increase or decrease the electron-electron repulsion? Explain.

17. Using the information from the graphs in **Model 3**, describe what happens to the atomic radius and ionization energy as you go across a row in the Periodic Table, e.g., from Li to Ne and Na to Ar?

18. Why do the trends that you identified in Key Question 17 occur? In your explanation use the increase in nuclear charge and the effectiveness of electron shielding of the nuclear charge by electrons in the same orbital as illustrated in **Model 2**.

19. Using the information from the graphs in **Model 3**, describe what happens to the atomic radius and ionization energy when going down a group in the Periodic Table, such as from Ne to Xe or Li to Rb.

20. Why do the trends that you identified in Key Question 19 occur? In your explanation, use the increase in nuclear charge, the effect of electron shielding of the nuclear charge by electrons in inner shells, and the size of the outer shell.

21. How can you use the Periodic Table and electron configurations to predict relative atomic radii and ionization energies for two atoms?

Exercises

4. Identify the larger of each pair, and explain why that one is larger.

   calcium or potassium

   K

   nitrogen or oxygen

   N

   sodium or potassium

   ~~Na~~ K

5. Identify which of the elements in each pair has the higher ionization energy, explaining why for each.

   Ba or Cs

   Ba

   Br or Kr

   Kr

   Si or C

   C

PROBLEMS

1. In the Periodic Table, ionization energies tend to increase across a period or row. For the ionization energies in the second period (Li to Ne), identify where the exceptions to this general trend occur. To help you explain these exceptions, use orbital box diagrams like those in **Model 1** and your knowledge of the differences in electron configurations between neighboring atoms.

2. Using the insight you gained from Problem 1, explain:

   a) Why the first ionization energy of sulfur is less than the first ionization energy of phosphorus.

   b) Why the first ionization energy of aluminum is less than the first ionization energy of magnesium.

3. Variations in electron affinities can be understood in the same way that variations in ionization energies are understood; except that, in the analysis of the effects of electron shielding and nuclear charge, one uses the electron configuration and the orbital box diagram for the anion rather than for the atom. Using this idea, identify the atom in each pair that has the greater electron affinity and explain why.

H or He

~~He~~ H

He or Li

He

B or C

C

Cl or Ar

Ar

4. Explain why the electron affinity of Be is very small, or zero, and the electron affinity of F is very large (328 kJ/mole).

F because
it has more electrons

5. Why do peaks in ionization energy occur for atoms with atomic numbers equal to 10, 18, and 36?

max number of electron

6. Elements with atomic numbers equal to 9, 17, and 35 have large electron affinities, but those elements with atomic numbers 10, 18, and 36 have very small, or zero, electron affinities. Why does an increase of one unit for the atomic number result in such a drastic change in electron affinity?

open spot in the p level
for an electron

Activity 08-1

The Chemical Bond

Why?

Chemical bonds form to lower the energy of the system of positive nuclei and negative electrons, which interact by the electrical force. This electrical force between pairs of atoms is called a chemical bond. The characteristics of chemical bonds lie at the heart of chemistry and are key to understanding and using chemistry.

Learning Objectives

- Identify how the energy of two atoms depends on the separation of the atoms
- Visualize the energy and force associated with the interaction of charged particles

Success Criteria

- Correct description of how the energy changes as two atoms approach each other
- Accurate interpretation of a graph of energy vs. the distance between two atoms to find the bond length as well as the bond energy (which is also called the bond strength)

Model: The Interaction of Two Hydrogen Atoms

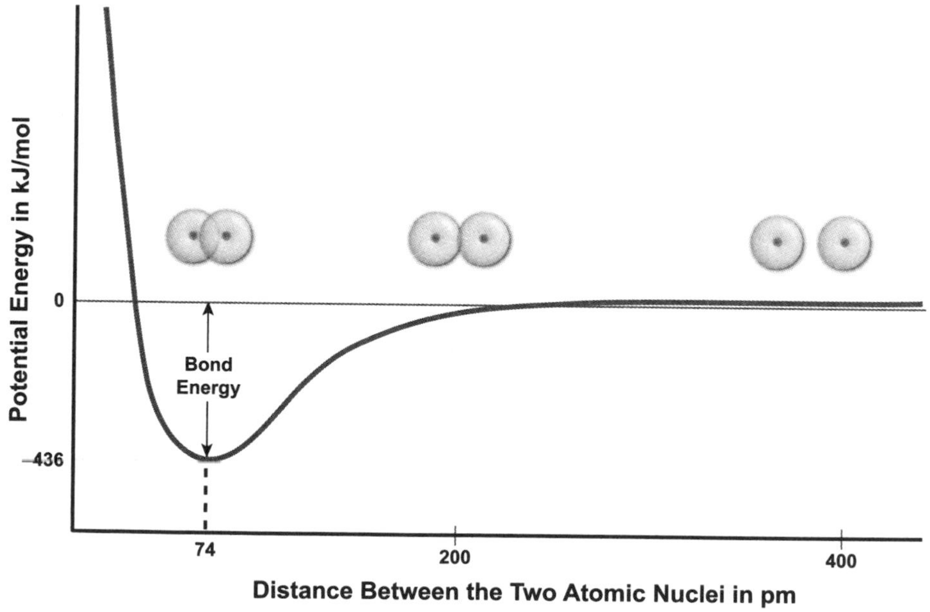

The energy of two atoms results from the attraction and repulsion of the charged electrons and nuclei and also from the kinetic energy due to the motion of the electrons and the nuclei. As two atoms approach each other from infinite separation, the two positive nuclei repel each other, the negative electrons of one are attracted by the positive nucleus of the other, and the electrons of one are repelled by the electrons of the other. At intermediate distances, the attractions dominate and the energy decreases to form a bond. The separation of the atoms when the energy is a minimum is called the *bond length*. Energy must be added to break the bond or separate the atoms; this energy is called the *bond energy*.

Suggestions contributed by Vicky Minderhout and colleagues at Seattle University

Key Questions

1. What quantity is plotted on each axis in the graph shown in the Model?

2. How far apart are the two hydrogen atoms in the Model where the energy is defined as being zero?

3. Why does the energy decrease as the two atoms come together? Respond using your own words.

4. Why does the energy increase as the two atoms come very close together?

5. Where on the abscissa in the graph is the distance corresponding to the bond length of the hydrogen molecule? Explain why you selected that particular value.

6. How much energy is needed to break the bond and produce hydrogen atoms from hydrogen molecules? How is this amount of energy indicated on the graph in the Model, and what is it called?

EXERCISE

Construct a diagram showing all the particles that make up a H_2 molecule, using arrows to show all the forces between them. Label each force as repulsive (R) or attractive (A).

Activity 08-2

Lewis Model of Electronic Structure

Why?

The Lewis model of molecular electronic structure describes how atoms bond to each other to form molecules. It determines the number of bonds formed between pairs of atoms in a molecule and the number of electrons that exist as lone or nonbonding pairs. This information makes it possible for you to predict the geometry of molecules (e.g., CO_2 is linear but SO_2 and H_2O are bent) and relative bond strengths and lengths.

Learning Objective

- Understand how to draw Lewis structures and interpret them in terms of molecular properties

Success Criteria

- Construct realistic Lewis structures
- Identify relative bond strengths and lengths from Lewis structures

Prerequisites

- **Activity 03-1:** *Molecular Representations*
- **Activity 07-4:** *Multi-electron Atoms, the Aufbau Principle, and the Periodic Table*

Information

Molecules exist because they are more stable than separated atoms. By "more stable," we mean that they have lower energy in the same way that a skateboarder at the bottom of a hill has less energy or is more stable than one at the top. The physical chemist G. N. Lewis recognized, from the very low chemical reactivity of the noble gases, that a configuration with eight electrons in a shell produces a very stable situation. He therefore proposed that molecules form so that atoms can transfer or share electrons and produce this very stable octet structure.

Lewis structures are used to model how the electrons are arranged in order to produce these stable eight-electron configurations. In these diagrams, dots are used to represent electrons; a line between atoms represents a single covalent bond formed by a pair of electrons; other dots represent nonbonding electrons; and charges are written to identify a formal distribution of charge.

The bonds show how the atoms in a molecule are connected to each other. A Lewis diagram does not show bond lengths, bond angles, the arrangement of atoms in three-dimensional space, or the actual charges on atoms. Some molecules require more than one Lewis structure to describe them. These multiple structures are called *resonance structures*. In some situations, atoms in Period 2 can have fewer than 8 electrons and atoms in Periods 3 and higher have more than 8 electrons.

Some atoms (e.g., C, N, O, and S) form double bonds, which are represented by double lines. Some atoms (e.g., C and N) form triple bonds, which are represented by triple lines. A double bond is stronger and shorter than single bond, and a triple bond is the strongest and shortest of the three.

How does one determine and draw a Lewis structure? First determine whether the molecule is ionic or covalent. If it is ionic, draw each ion separately. For covalent molecules and polyatomic ions, follow the methodology given in **Model 1**.

Model 1: Methodology for Constructing Lewis Structures

HCl – A Simple Example

Methodology		Example: Single Bond
Step 1:	Count the valence electrons from all the atoms. Add electrons for negative ions, subtract electrons for positive ions. The number of valence electrons can be determined from the atom's position in the Periodic Table.	For hydrochloric acid, H has 1 and Cl has 7 for a total of 8 valence electrons.
Step 2:	Assemble the bonding framework. Decide which atoms are connected to each other, and use a pair of electrons, represented by a line, to form a bond between each pair of atoms that are bonded together.	**H—Cl**
Step 3:	Arrange the remaining electrons so that each atom has 8 electrons around it (the octet rule). If necessary, place additional pairs of electrons between the atoms to form additional bonds.	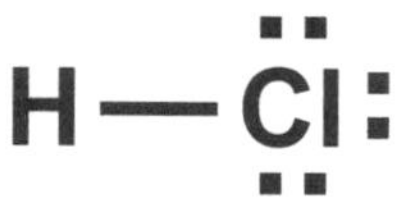
Step 4:	Check for exceptions to the octet rule. For H, only 2 electrons are needed (the duet rule). Be and B are often electron deficient and may have only 4 or 6 electrons. Atoms in the second period cannot have more than 8 electrons. The third and higher period elements sometimes have more than 8 electrons.	Cl satisfies the octet rule. H satisfies the duet rule.
Step 5:	Determine the formal charges (FC) on the atoms. FC = number of atomic valence electrons – number of lone pair electrons – 0.5(number of shared electrons). Evaluate whether the formal charges (FC) on the atoms are reasonable. The structure is reasonable if the charges are zero or small, and the negative charges reside on the most electronegative atoms. Other structures with larger formal charges are higher energy and are not representative of the lowest energy structure of the molecule.	FC(H) = 1 – 0 – 0.5 (2) = 0 FC(Cl) = 7 – 6 – 0.5 (2) = 0 Reasonable because FC is zero.
Step 6:	Draw resonance structures.	none in this case

Key Questions

1. How can you determine the number of valence electrons that an atom has?

2. a) How many valence electrons does H have and how many more does it need to fill the first shell?

   b) How many valence electrons does Cl have and how many more does it need to achieve a noble gas configuration?

3. From the Lewis perspective, why do hydrogen and chlorine combine to form the HCl molecule?

4. How is formal charge determined?

5. How is formal charge used to identify unreasonable Lewis structures?

EXERCISES

1. Write the number of valence electrons for H and for F. Draw the Lewis structure for HF.

2. Write the number of valence electrons for C. Determine how many more electrons C needs to form a molecule, and draw the Lewis structure for a molecule composed of C and H.

3. Write the number of valence electrons for S. Determine how many more electrons S needs to form a molecule, and draw the Lewis structure for a molecule composed of S and H.

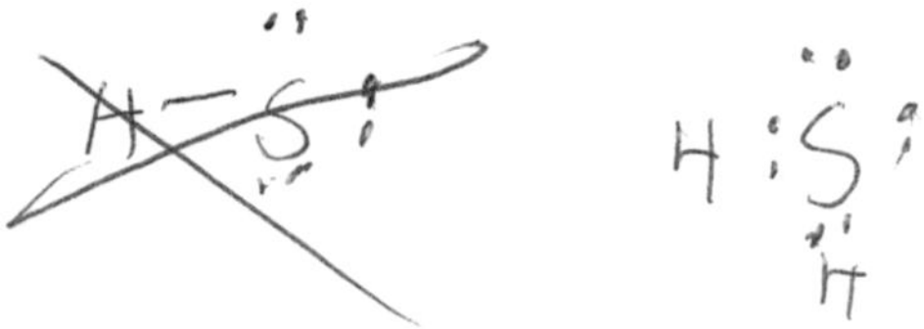

4. Draw Lewis structures for H_2O, NH_3, PCl_3, C_2H_6, and NaCl. (NaCl is ionic, so draw the two ions separately, with no line connecting them, and indicate the charge on the ion with a + or – sign.)

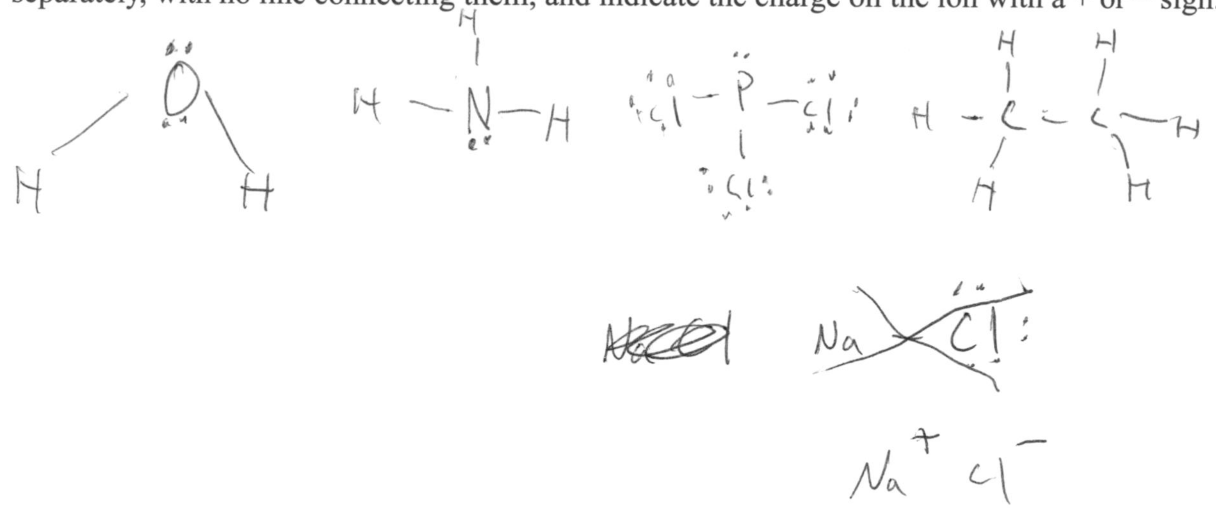

Table 1

CO_2 – Need for Double Bonds

Methodology	Example: Multiple Bonds
Step 1: Count the number of valence electrons from all the atoms.	For carbon dioxide, C has 4 valence electrons and O has 6 valence electrons for a total of 16 valence electrons.
Step 2: Assemble the bonding framework.	**O — C — O**
Step 3: Arrange the remaining electrons so that each atom has 8 electrons around it (the octet rule). If necessary, place additional pairs of electrons between the atoms to form additional bonds to satisfy the octet rule.	**:Ö = C = Ö:**
Step 4: Check for exceptions to the octet rule.	No exceptions present.
Step 5: Evaluate whether the formal charges on the atoms are reasonable.	FC(C) = 4 – 0 – 0.5 (8) = 0 FC(O) = 6 – 4 – 0.5 (4) = 0 Reasonable because FC is zero.
Step 6: Draw resonance structures.	none in this case

Key Questions

6. How many valence electrons are there in the carbon dioxide molecule?

7. Why is it sometimes necessary in constructing Lewis structures to put double or even triple bonds between atoms?

Exercises

5. Draw the Lewis structures for O_2, N_2, C_2H_4, and C_2H_2.

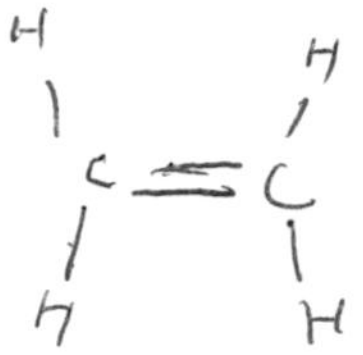

ICl_4^- – Exception to the Octet Rule

Table 2

Methodology	Example: Third Period Element
Step 1: Count the number of valence electrons from all the atoms, and, because it is a –1 anion, add one.	For ICl_4^-, there are 36 valence electrons ($5 \times 7 + 1$)
Step 2: Assemble the bonding framework.	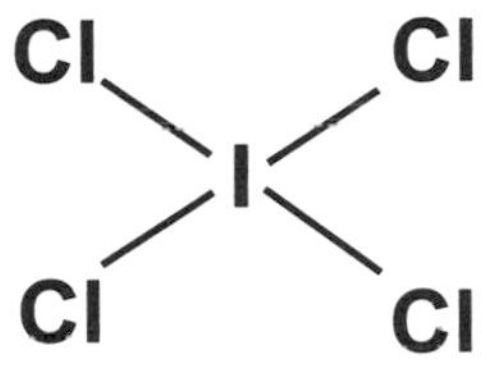
Step 3: Arrange the remaining electrons so that each atom has 8 electrons around it.	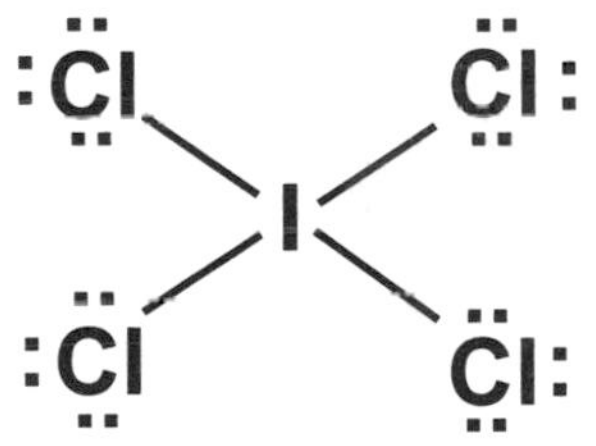
Step 4: Check for exceptions to the octet rule. After satisfying the octet rule for each atom, 4 electrons remain. These are placed as nonbonding electrons on the fifth period element, iodine.	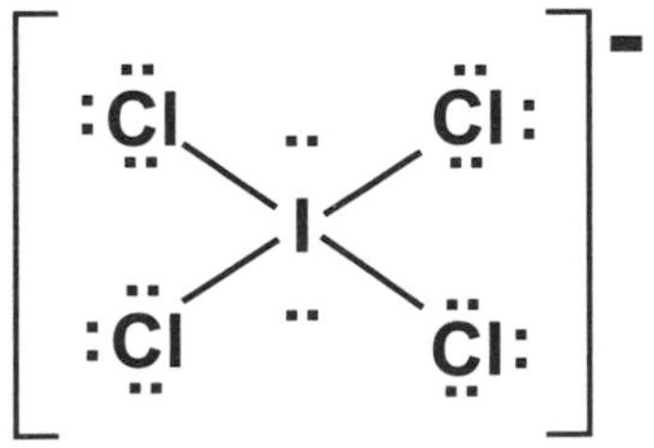
Step 5: Evaluate whether the formal charges on the atoms are reasonable. *Note:* *The negative formal charge is on I, which is less electronegative than Cl. This result is correct and is an exception to the general rule given earlier.*	FC(I) = 7 – 8 = –1 FC(Cl) = 7 – 7 = 0
Step 6: Draw resonance structures.	none in this case

Key Questions

8. How does the example of ICl_4^- differ from the examples of HCl and CO_2?

Exercise

5 + 35 = 40

6. Draw the Lewis structure for PCl_5.

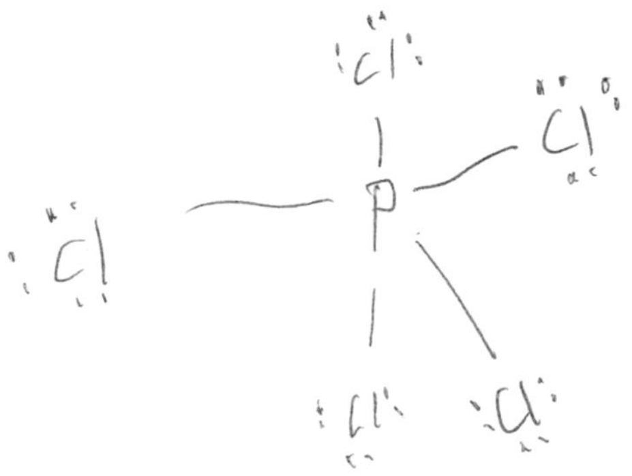

NO_2^- – The Need for Multiple Lewis Structures

Table 3

Methodology		Example: Resonance
Step 1:	Count the number of valence electrons from all the atoms, and, because it is a –1 ion, add one.	For NO_2^- there are 18 valence electrons (5 + (2 × 6) + 1)
Step 2:	Assemble the bonding framework.	**O — N — O**
Step 3:	Arrange the remaining electrons so that each atom has 8 electrons around it.	$[:\ddot{\underset{..}{O}}-\ddot{N}=\ddot{\underset{..}{O}}]^-$
Step 4:	Check for exceptions to the octet rule.	none
Step 5:	Evaluate whether the formal charges on the atoms are reasonable.	FC(N) = 5 – 5 = 0 FC(O1) = 6 – 7 = –1 FC(O2) = 6 – 6 = 0
Step 6:	In Step 3 the double bond could just as well have been drawn on the left side of the nitrogen. This second structure is called a resonance structure. In fact, experimental measurements show that both N-O bonds are the same with a bond energy and a bond length that are intermediate between a typical N-O single bond and a typical N=O double bond. Therefore, the second structure is needed to adequately describe the bonding. Draw the resonance structure to show that both bonds are equivalent.	$[\ddot{\underset{..}{O}}=\ddot{N}-\ddot{\underset{..}{O}}:]^-$

KEY QUESTIONS

9. Why is it not possible to describe NO_2^- by a single Lewis structure?

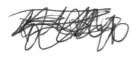

EXERCISE

7. Draw the Lewis structure for ozone O_3. Include the formal charge in parentheses above each atom if the formal charge differs from zero, e.g., (+1) or (–1).

MODEL 2: BOND STRENGTHS AND LENGTHS

Bond	Average Dissociation Energy in kJ/mole	Average Length in pm
C–C	345	154
C=C	615	133
C≡C	835	120

KEY QUESTIONS

10. The energy it takes to dissociate or break a bond is a measure of the bond strength. In view of the data in **Model 2**, how does the Lewis structure help you identify the strongest bonds in a molecule?

11. In view of the data in **Model 2**, how does the Lewis structure help you identify the shortest bonds in a molecule?

Exercise

8. Identify which C–O bond in acetic acid is the shortest and strongest.

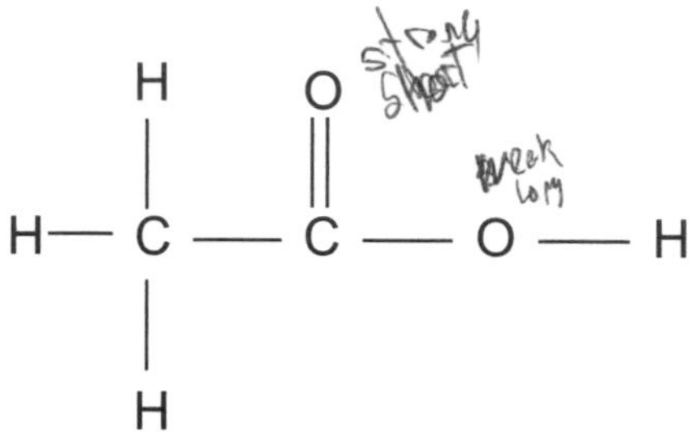

Model 3: Which Lewis Structure is Better?

The two structures below represent the thiocyanate ion. Which one has a lower energy and is therefore the better description of the lowest energy state of this ion?

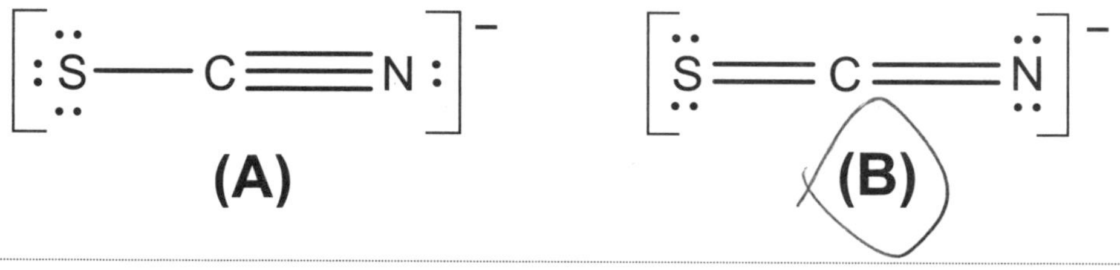

Key Questions

12. What are the formal charges on the atoms in the thiocyanate ion in **Model 3**? Write the formal charge in parentheses above each atom in these structures.

13. Based on the criteria developed in Key Question 5, which structure for thiocyanate (in **Model 3**) has the lower energy and is, therefore, the better description of the lowest energy state of this ion?

EXERCISES

9. In view of the formal charges, why is O = C = O a better representation of carbon dioxide than O − C ≡ O ?

O=C=O gives you zeros

INFORMATION

In assembling the bonding framework of a molecule, you may find it difficult to identify which atoms are bonded to each other. Sometimes this information cannot be deduced from first principles: you simply need to know the molecular structure. Here are some guidelines that are often helpful.

- It is useful to think in terms of *outer atoms* and *inner atoms*. An *outer atom* bonds to only one other atom, while an *inner atom* bonds to more than one other atom.
- Hydrogen atoms are always outer atoms because they can form only one bond.
- Outer atoms other than hydrogen are usually the ones with the highest electronegativities.
- The bonding framework is often indicated by the order in which the atoms are written in a molecular formula. For example, in OCS the carbon atom is the inner atom.
- Parentheses are often used in a molecular formula to indicate the bonding framework. For example in $(CH_3)_2CO$, the hydrogen atoms are bonded to carbon to form two methyl radicals, and oxygen is an outer atom bonded to an inner carbon atom.
- Multiple atoms of the same element are usually the outer atoms around a single atom of another element. For example, in PF_6, phosphorus is the inner atom.
- Sometimes it is helpful to know the chemical properties of a molecular formula. For example, in HNO_3 the hydrogen atom might be bonded to the nitrogen or to oxygen. If you know that nitric acid is an oxyacid, you can locate the hydrogen as bonded to an oxygen atom.
- Finally, the most likely structure is the one which has the most reasonable formal charges on the atoms. By "reasonable," we mean that the formal charges should be small or zero, and the negative charges should be located on the most electronegative atoms.

EXERCISES

10. Draw Lewis structures for the following molecules. A Lewis structure includes the formal charge on each atom if it differs from zero and any resonance structures that are significant. To be significant, a resonance structure does not increase the formal charge on any of the atoms. Indicate the formal charge on an atom by writing it in parentheses next to the atom, e.g., F (–1).

SeF_2 | I_3^- | OCS

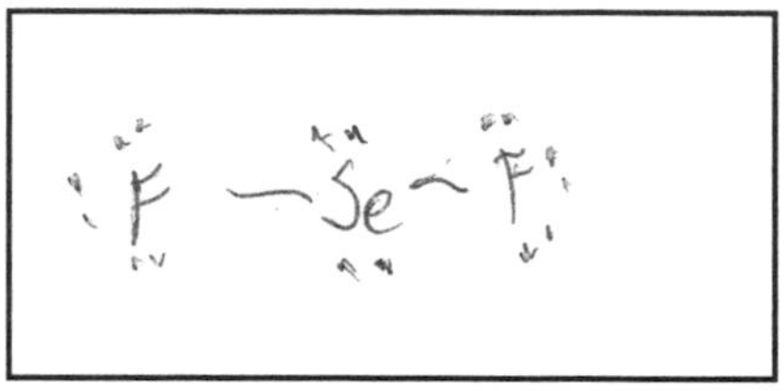
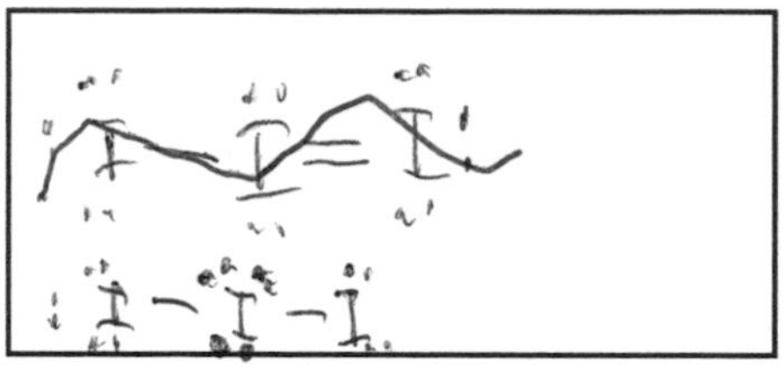
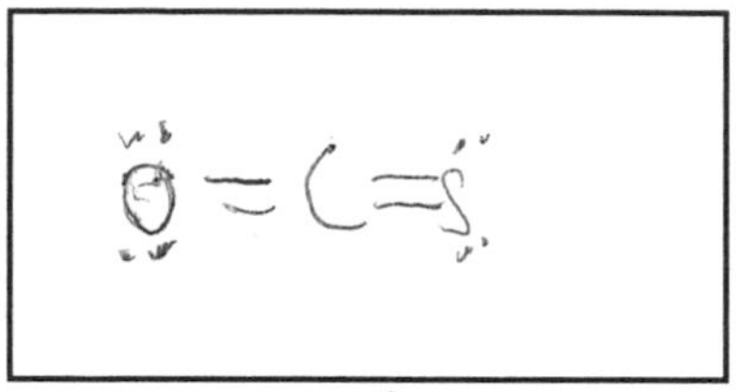

HNO_3 | $SiCl_4$

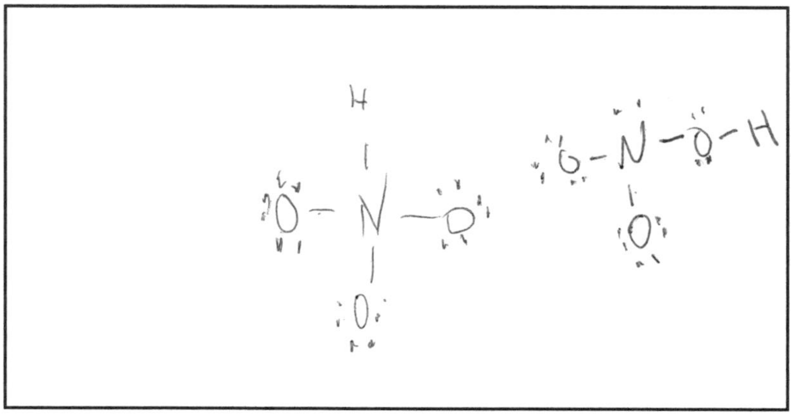
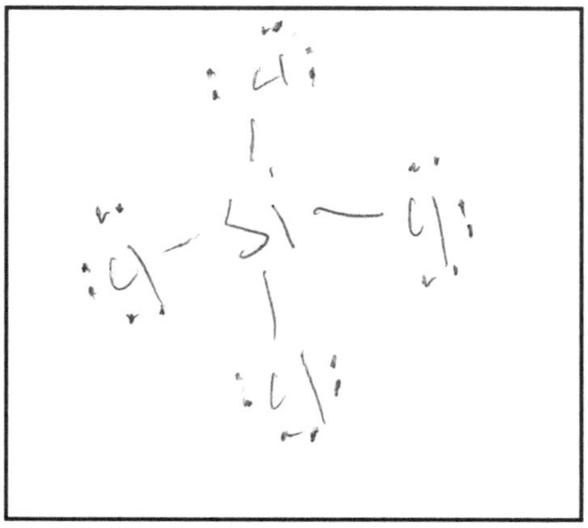

11. Use Lewis structures to arrange the following compounds in order of increasing carbon-carbon bond ***strength***. Explain.

C_2H_6 (ethane) | C_2H_4 (ethylene) | C_2H_2 (acetylene)

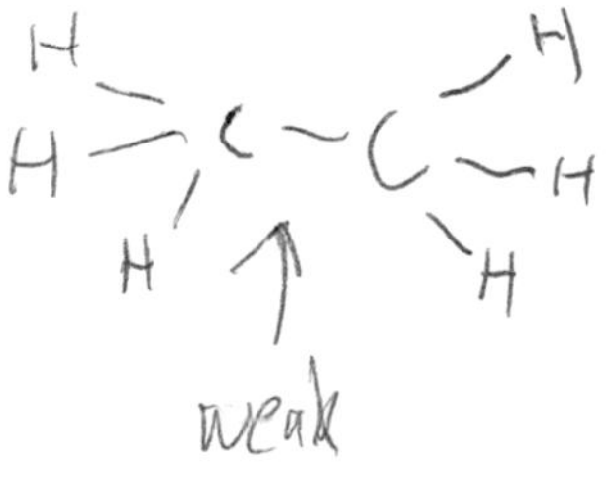

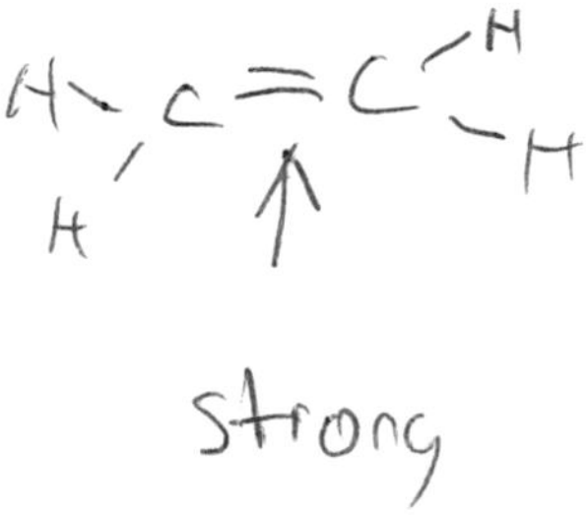

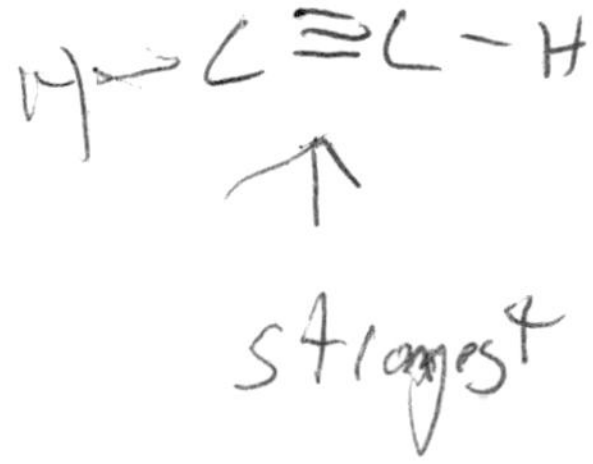

12. Use Lewis structures to arrange the following compounds in order of increasing carbon-carbon bond ***length***. Explain.

C_2H_6 (ethane) | C_2H_4 (ethylene) | C_2H_2 (acetylene)

13. Explain which N–N bond is more stable: the one in nitrogen (N_2) or the one in hydrazine (N_2H_4).

14. Explain which C–O bond is shorter: the one in methanol (CH_3OH) or the one in formaldehyde (CH_2O).

PROBLEMS

1. Use Lewis structures to identify which of the following compounds you would most likely be successful at synthesizing: SiF_4, OF_4, SF_6, or OF_6. Explain.

2. Two Lewis structures are needed to describe the bonding in formamide, $HCONH_2$. Write these two resonance structures. One should have no formal charges on the atoms, and the other will have a formal charge of +1 on N and –1 on O. Of the structures you have drawn, which do you expect to be the better description of formamide?

3. For the two resonance structures of formamide in Problem 2, explain why each of the following statements is either correct or incorrect.
   a) The molecule is ***not*** oscillating back-and-forth between the two structures; the molecule is, instead, an average or superposition of the two structures.

   b) The expected CO bond length is between that for a normal CO double bond and that for a normal CO single bond.

   c) The nitrogen in the molecule has a lower electron density associated with it than is found for the free nitrogen atom.

   d) Both resonance structures have the same energy.

Reflection

Develop a checklist that you can use to ensure that you have written a correct Lewis structure.

Activity 08-3

Valence Bond Model for Covalent Bonds

Why?

Covalent bonds are formed by atoms sharing pairs of electrons (as represented by Lewis structures), but Lewis structures do not provide any information about the orbitals occupied by the bonding and lone pair electrons. The valence bond model addresses this deficiency by describing a covalent bond as formed from the overlap of atomic orbitals on each of the two atoms that are joined by the bond. The overlap region between the two atoms builds up electron density that attracts the positively charged atomic nuclei and holds the atoms together. This model accounts for the fact that not all bond strengths are equal in terms of the overlap of the orbitals; the greater the overlap, the stronger the bond.

Learning Objective

- Characterize sigma and pi bonds in terms of overlapping atomic orbitals

Success Criteria

- Correctly describe sigma and pi bonds in terms of overlapping atomic orbitals
- Correctly describe single, double, and triple bonds in terms of sigma and pi bonds

Prerequisites

- **Activity 07-3:** *The Description of Electrons in Atoms*
- **Activity 08-2:** *Lewis Model of Electronic Structure*

Model: Representation of Covalent Bonds by Overlapping Atomic Orbitals

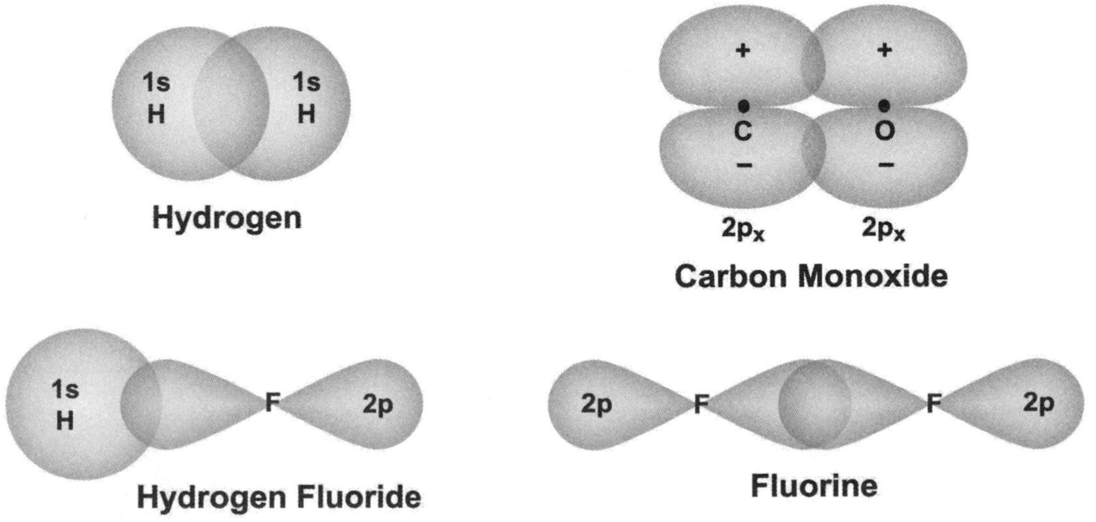

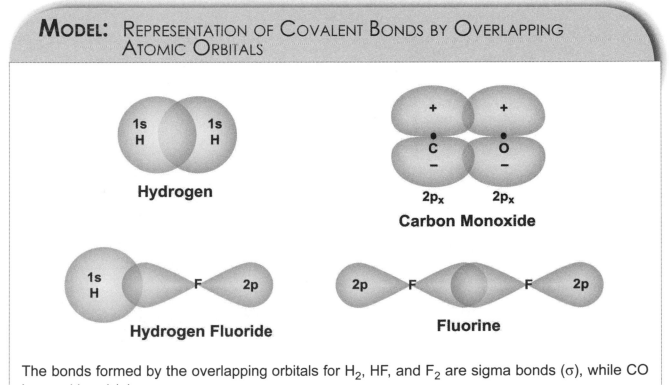

The bonds formed by the overlapping orbitals for H_2, HF, and F_2 are sigma bonds (σ), while CO has a pi bond (π).

Suggestions contributed by Vicky Minderhout and colleagues at Seattle University

Key Questions

1. Based on the Model, what types of atomic orbitals are overlapping in each case?

   hydrogen:

   hydrogen fluoride:

   carbon monoxide:

   fluorine:

2. What is the characteristic geometry of sigma and pi bonds, *end-to-end* or *side-to-side?*

   sigma bonds:

   pi bonds:

3. What is the characteristic shape of sigma and pi bonds with respect to the line connecting the two nuclei? (This line is also called the *internuclear axis* or the *bond axis*.)

   sigma bonds:

   pi bonds:

4. Where is the electron density in each of these bonds with respect to the internuclear axis?

   sigma bonds:

   pi bonds:

5. If a bond is a single bond as in H_2, HF, or F_2, what type is it (sigma or pi)?

Exercises

1. Draw pictures of the electron density patterns in a sigma bond and in a pi bond.

2. Carbon monoxide has a triple bond consisting of one sigma bond and two pi bonds. Draw a diagram to show how the two pi bonds can be formed from overlapping the p_x atomic orbitals of carbon and oxygen for one pi bond and overlapping p_y atomic orbitals for the other pi bond.

ACTIVITY 08-4

Molecular Orbital Theory

WHY?

Just as atomic orbitals describe electrons in atoms, molecular orbitals describe electrons in molecules. Molecular orbitals account for the delocalization of electrons in the molecule and provide an alternative description to the Lewis model, which views the electrons as being localized in bonds between pairs of atoms. You can use the molecular orbital model to determine bond order, and, consequently, predict the strength, energy, and length of bonds, and magnetic properties of molecules.

LEARNING OBJECTIVES

- Identify how molecular orbitals can be formed from atomic orbitals
- Understand how the electron occupation of molecular orbitals is related to bond order, bond strength, bond energy, bond length, and magnetic properties

SUCCESS CRITERIA

- Construct and identify molecular orbitals
- Analyze bond order, bond strength, bond energy, bond length, and magnetic properties in terms of the occupation of molecular orbitals

PREREQUISITES

- **Activity 07-5:** *Periodic Trends in Atomic Properties*
- **Activity 08-2:** *Lewis Model of Electronic Structure*

INFORMATION

Molecular orbitals can be written as sums of atomic orbitals multiplied by a coefficient, which often is +1 or –1. Some of these orbitals are *bonding*, some *nonbonding*, and some *antibonding*.

A *bonding orbital* has a high electron density between two atomic nuclei. This negative electron charge attracts the positive nuclei and holds them together.

A *nonbonding orbital* positions electrons away from the bonds and does not contribute to bonding, e.g., nonbonded or lone pairs in the Lewis model would be in nonbonding orbitals in the molecular orbital model.

An *antibonding orbital* has little electron density between the two atomic nuclei, allowing the nuclei to repel each other. This repulsive effect of electrons which are in antibonding orbitals cancels the attractive effect of electrons in bonding orbitals, so molecules with equal occupations of bonding and antibonding orbitals are not stable.

The *bond order* is given by half the number of bonding electrons minus half the number of antibonding electrons: $BO = \frac{1}{2}(n_{bonding} - n_{antibonding})$ since the bonding electrons attract the atomic nuclei together, and antibonding electrons destabilize this bond. The bond order correlates with bond strength, bond energy, and bond length. The higher the bond order, the stronger the bond; the higher the bond energy, the shorter the bond length.

Model 1: Valence Orbitals of Diatomic Molecules

Figure 1

Notation

atomic orbitals on atom labeled A: $2s(A)$, $2p_x(A)$, $2p_y(A)$, $2p_z(A)$

atomic orbitals on atom labeled B: $2s(B)$, $2p_x(B)$, $2p_y(B)$, $2p_z(B)$

σ designates a sigma molecular orbital, which has a maximum electron density along a line connecting the two nuclei. This line is called the *internuclear axis*.

π designates a pi molecular orbital, which has a maximum electron density above and below the internuclear axis

$*$ designates an antibonding molecular orbital, which has a region of very low electron density between the two nuclei

Figure 2

Valence Molecular Orbitals Formed from Atomic Orbitals

$\sigma_{2s} = 2s\ (A) + 2s\ (B)$
$\sigma_{2s}* = 2s\ (A) - 2s\ (B)$

$\sigma_{2p_z} = 2p_z\ (A) - 2p_z\ (B)$
$\sigma_{2p_z}* = 2p_z\ (A) + 2p_z\ (B)$

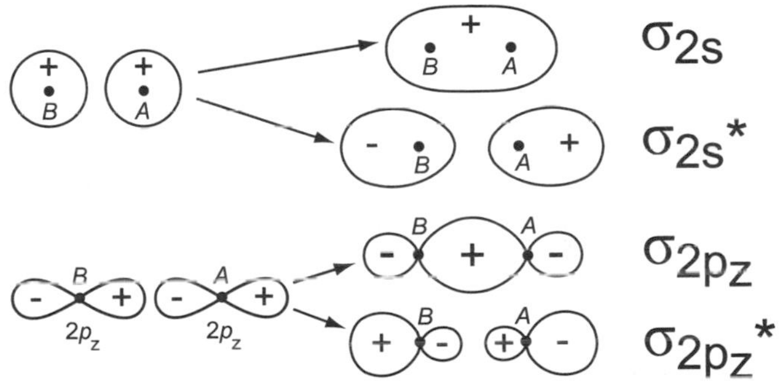

$\pi_{2p_x} = 2p_x\ (A) + 2p_x\ (B)$
$\pi_{2p_x}* = 2p_x\ (A) - 2p_x\ (B)$

The π_{2p_x} orbitals are not shown, because they look just like the π_{2p_y} orbitals but in the x-direction.

$\pi_{2p_y} = 2p_y\ (A) + 2p_y\ (B)$
$\pi_{2p_y}* = 2p_y\ (A) - 2p_y\ (B)$

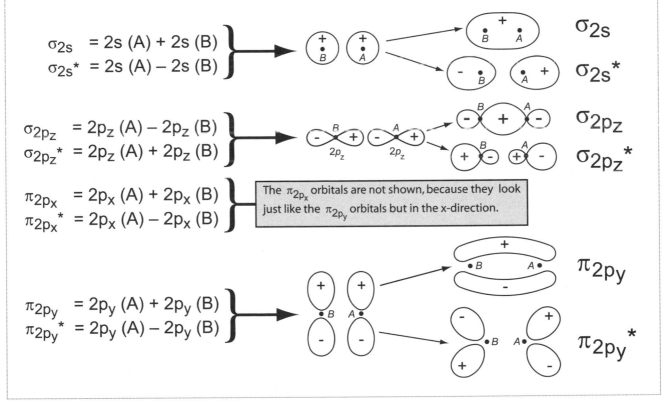

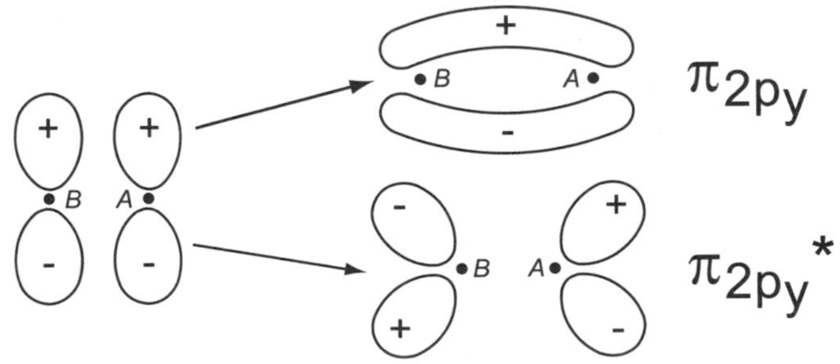

Key Questions

1. How many molecular orbitals are produced by combining two atomic orbitals as illustrated in **Model 1**?

2. If two s atomic orbitals from different atoms are combined, what are the names of the molecular orbitals that are produced and where are the regions of high and low electron density for these orbitals?

3. If two p_z atomic orbitals from different atoms are combined, what are the names of the molecular orbitals that are produced? Where are the regions of high and low electron density for these orbitals?

4. If two p_y atomic orbitals from different atoms are combined, what are the names of the molecular orbitals that are produced? Where are the regions of high and low electron density for these orbitals?

5. According to **Model 1**, what are the differences between a bonding and an antibonding molecular orbital that apply to all three pairs of bonding and antibonding orbitals shown in **Model 1**?

6. In terms of the relationship to the internuclear axis, what is the difference between a π and a σ molecular orbital?

7. What are the similarities between a σ_{2s} and a σ_{2pz} molecular orbitals?

8. What are the differences between a σ_{2s}^* and a σ_{2pz}^* molecular orbitals?

Exercises

1. Write the π_{2px} and π_{2px}^* molecular orbitals in terms of the $2p_x$ atomic orbitals.

2. Draw diagrams similar to those in **Model 1** to show how the π_{2px} and $\pi_{2px}{}^*$ molecular orbitals are formed from the $2p_x$ atomic orbitals.

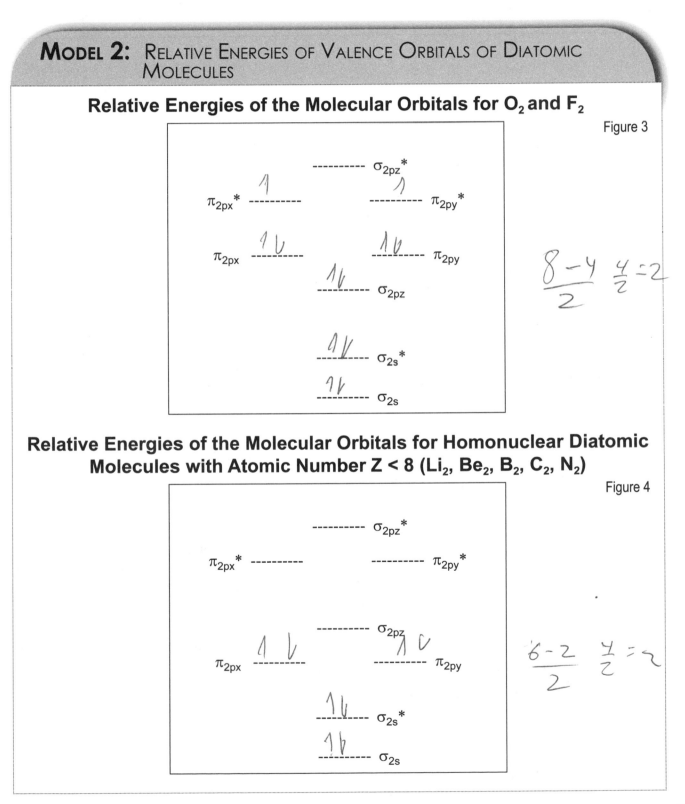

Model 2: Relative Energies of Valence Orbitals of Diatomic Molecules

Relative Energies of the Molecular Orbitals for O_2 and F_2

Figure 3

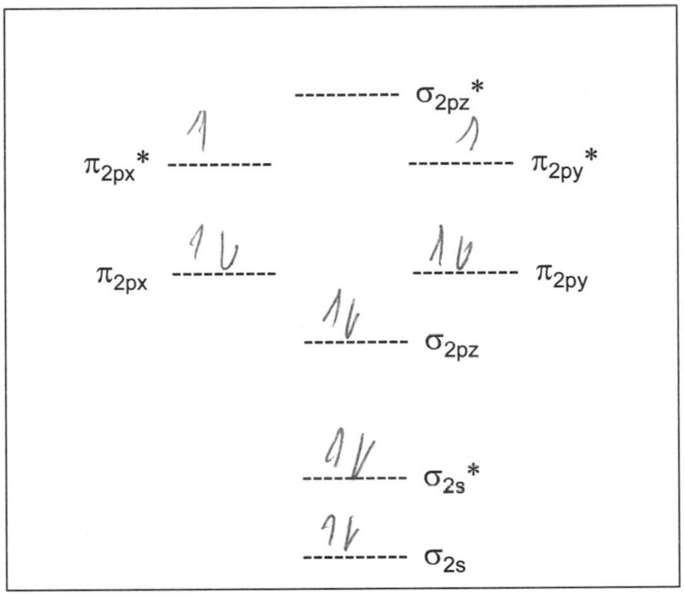

$\frac{8-4}{2}$ $\frac{4}{2}=2$

Relative Energies of the Molecular Orbitals for Homonuclear Diatomic Molecules with Atomic Number Z < 8 (Li_2, Be_2, B_2, C_2, N_2)

Figure 4

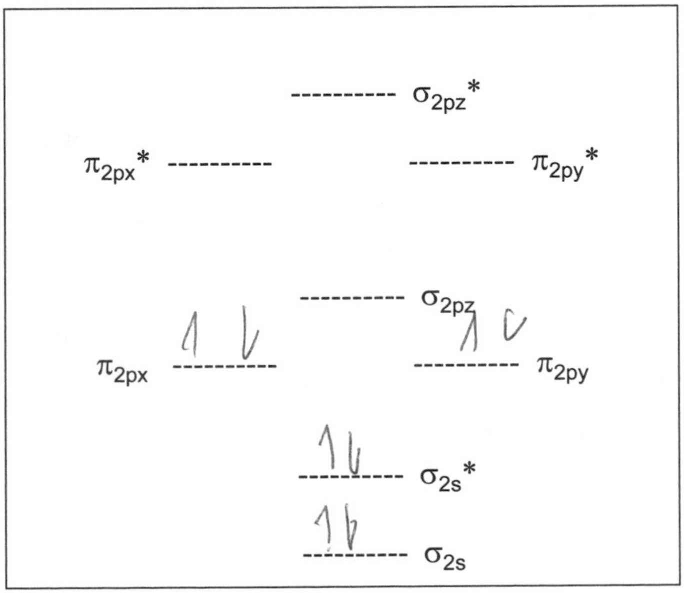

$\frac{6-2}{2}$ $\frac{4}{2}=2$

Key Questions

9. When the atomic number $Z < 8$, which molecular orbital for homonuclear diatomic molecules is moved up in energy relative to the molecular orbitals for O_2 and F_2?

10. In view of the geometry or shape of diatomic molecules, why do you think that the π_{2px} and π_{2py} orbitals have the same energy?

11. What are six insights your team gained about molecular orbitals by examining the two models?

Exercises

3. For the cases of C_2 and O_2, place the electrons in the energy-level diagrams in **Model 2**. Represent electrons with up and down arrows, and apply the Aufbau Principle, the Pauli Exclusion Principle, and Hund's Rule, as discussed in Activity 07-5 on Periodic Trends in Atomic Properties.

4. Using the insight you have gained from the energy level diagrams that you constructed in Exercise 3, identify which of the following statements are correct for C_2 and O_2.

   a) Both molecules have unpaired electrons.

   No

   b) Only oxygen has unpaired electrons.

   yes

c) Both molecules have a bond order of 2.

true

d) The bond strength and bond energy are predicted, from the occupation of the orbitals, to be larger for O_2 than for C_2.

e) Both are homonuclear diatomic molecules.

f) Oxygen is paramagnetic. (See the **Information** section which follows.)

5. Some heteronuclear diatomic molecules have energy levels like those in Figure 3, while others have energy levels like those in Figure 4, e.g, CN, CN^-, and CN^+

   a) Write the molecular orbital electron configurations for these three species. For example, the electron configuration for H_2 is $(\sigma_{1s})^2$.

   b) Placc these species in order of increasing bond order, increasing bond length, increasing bond energy, and increasing vibrational frequency. (The higher the bond order and strength, the higher the vibrational frequency.)

6. Identify the characteristic that determines whether a molecule is paramagnetic or diamagnetic.

7. Label the following species as paramagnetic or diamagnetic.

O_2	______________	N_2	______________	C_2	______________
CN	______________	CN^-	______________	CN^+	______________

Information

Electrons behave as tiny bar magnets because they have a magnetic moment. If electrons are paired in molecular orbitals, the magnetic moments of the two electrons cancel each other out because the magnets point in opposite directions. Such a substance is *diamagnetic*. Unpaired electrons produce a net magnetic moment, and the molecule then is said to be *paramagnetic*.

Got It!

1. Identify which of the following statements are correct when molecular orbitals (MOs) are formed from two 1s atomic orbitals, each centered on different H-atom nuclei (protons).
   a) Two bonding MOs are formed.

   b) Two antibonding MOs are formed.

   c) One bonding and one antibonding MO are formed, with the bonding MO having the lower energy.

   d) One bonding and one antibonding MO are formed, with the antibonding MO having the lower energy.

   e) If the MOs are occupied by electrons with the **same** value of m_S, both electrons will be in the bonding MO, as required by the Pauli Exclusion Principle and Hund's Rule.

Activity 09-1

Valence Shell Electron Pair Repulsion Model

Why?

Molecules adopt a shape that minimizes their energy. In many cases, it is possible to predict the geometry of a molecule simply by considering the repulsive energy of electron pairs. You can use this valence shell electron pair repulsion model (VSEPR) to predict shapes and determine whether or not a molecule is polar. Scientists commonly use this model when they need to predict or estimate the shape of a molecule.

Learning Objective

- Understand how molecular shape is predicted from the Lewis structure

Success Criterion

- Accuracy in predicting molecular shapes

Prerequisite

- **Activity 08-2:** *Lewis Model of Electronic Structure*

Information

The terms *Lewis structure*, *electronic structure*, *electron arrangement*, and *electron geometry* are used to describe how the bonding and nonbonding electron pairs are positioned in a molecule. The terms *molecular shape*, *molecular structure*, and *molecular geometry* are used to describe how the atoms are positioned relative to each other in a molecule.

Model 1: Methodology for Determining Molecular Geometries (Shapes or Structures) from the VSEPR Model

Methodology	Example
Step 1: Draw the Lewis electronic structure.	*For sulfur dioxide:* 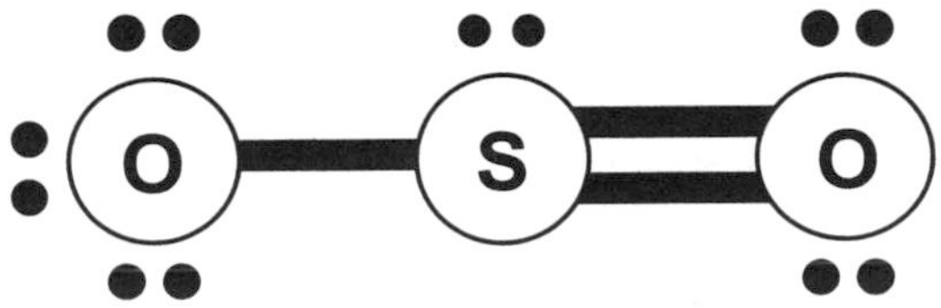
Step 2: Count the number of bonds and nonbonding electron pairs around the central atom.	*1 single + 1 double bond + 1 nonbonding pair = 3. This number is called the steric number.*

continued on following page

Model 1: Methodology for Determining Molecular Geometries (Shapes or Structures) from the VSEPR Model (Con't)

Methodology	Example
Step 3: Molecules take a shape that minimizes their energy. Arrange the bonds and nonbonding electron pairs to maximize their separation, which minimizes the electron-electron repulsion energy.	*A steric number of 3 in step 2 means a trigonal planar electronic structure minimizes the energy:* 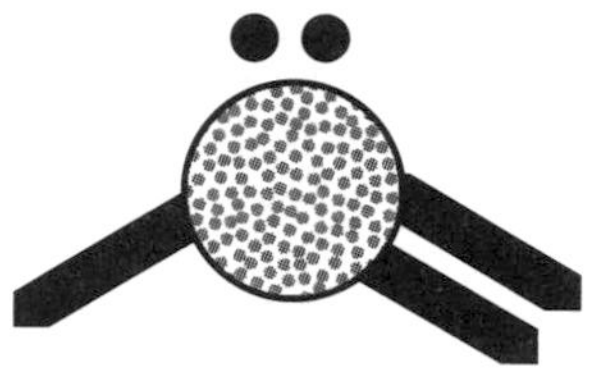
Step 4: Add the atoms in a way that is consistent with how the electrons are shared, and put the nonbonding electron pairs as far apart as possible.	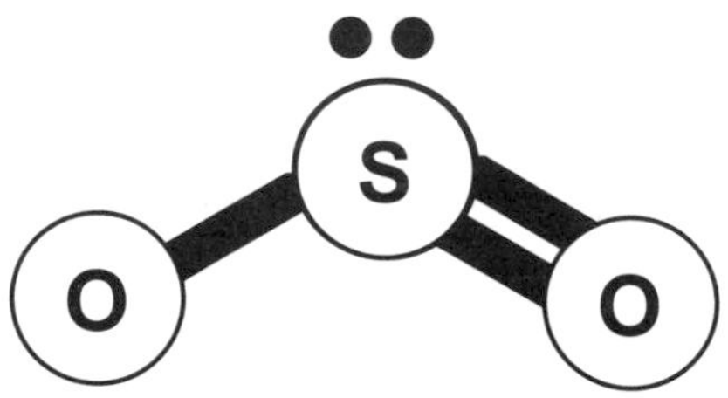
Step 5: Determine the molecular shape from the position of the atoms.	*The atoms are arranged in a nonlinear or bent shape.*

Key Questions

1. In Step 1 in the preceding methodology, how do you determine the Lewis electronic structure?

2. Why are bonds and nonbonding electron pairs (aka: *lone pairs*) spaced as far apart as possible in the structure?

3. According to Step 4 in the methodology, if you have two lone pairs and bonds to four atoms around a central atom, would you position the lone pairs at 90° or 180° to each other? Explain.

4. How would you describe the geometrical arrangement of the bonds and lone pairs around sulfur in sulfur dioxide?

5. How would you describe the shape of sulfur dioxide?

6. Some triatomic molecules are linear. What feature of SO_2 leads to the bent geometry?

7. What three insights has your team gained about the shape of molecules by examining the model and responding to the key questions?

EXERCISES

1. Complete the illustrations in the following table to show the arrangement of bonds and electron lone pairs that minimizes the energy in each case. Your illustration represents the *Lewis electronic structure* of the molecule. The number of bonds and lone pairs is called the *steric number*.

Table 2

Number of Bonds and Lone Pairs	Lewis Electronic Structure	Illustration of the Electronic Structure
2	linear	
3	trigonal planar	
4	tetrahedral	
5	trigonal bipyramidal	
6	octahedral	

2. All of the possible molecular shapes for atoms arranged around a central atom are shown in the *Illustration* column of Table 3. Each of these shapes is exemplified by one molecule from the following list.

O_3 I_3^- IF_6^+ SbF_5 $COCl_2$ SeO_3^{2-} SiF_4 KrF_4 SF_4 ICl_3 BrF_5

Use the VSEPR model to predict the shape of each molecule and draw its Lewis structure in the left column of the row corresponding to its structure in the Table.

Table 3

Structure/Example	**Illustration**
I_5^-	
O_3	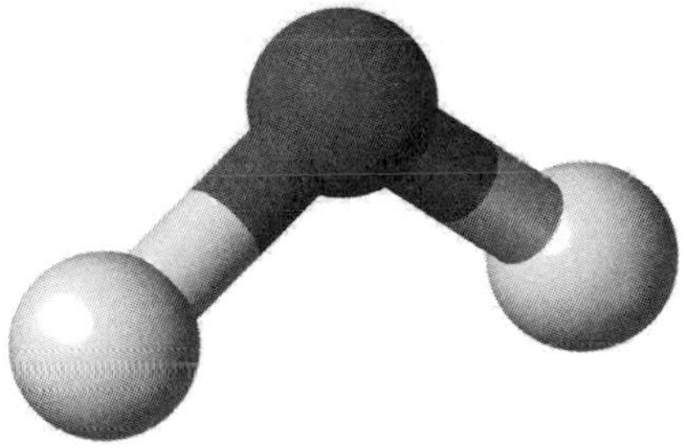
$COCl_2$	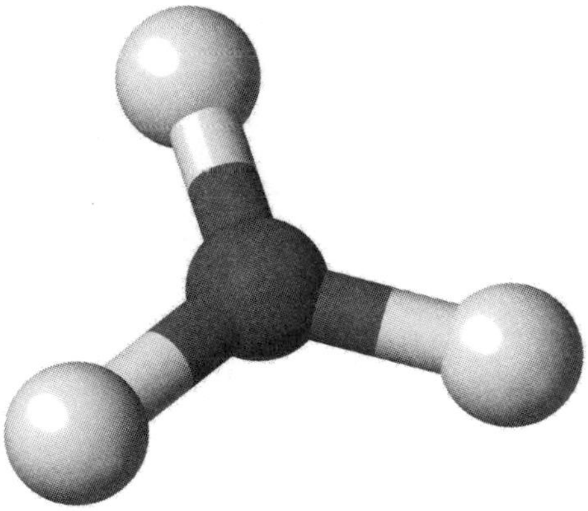
KrF_4	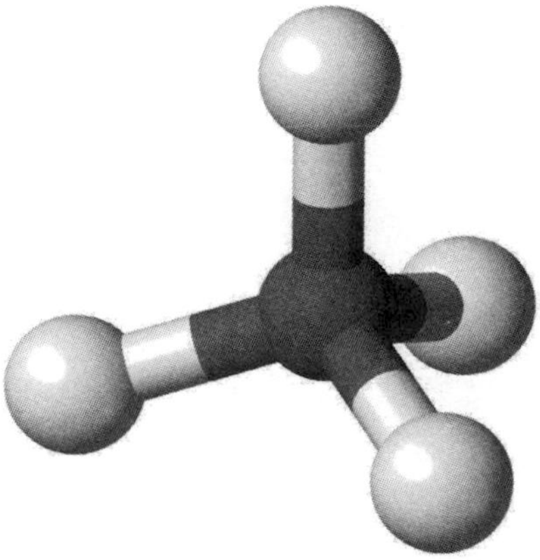

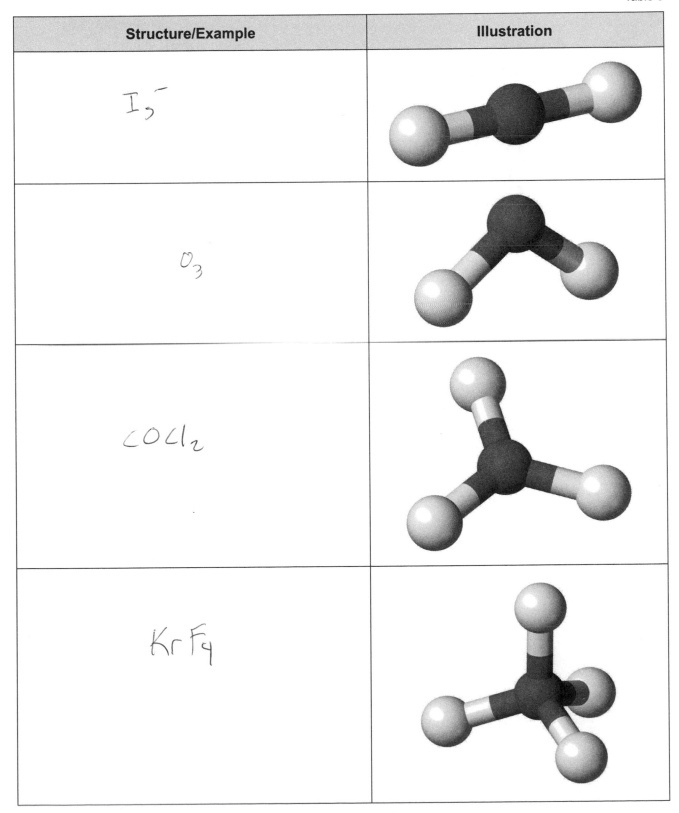

Structure/Example	Illustration
SbF_5	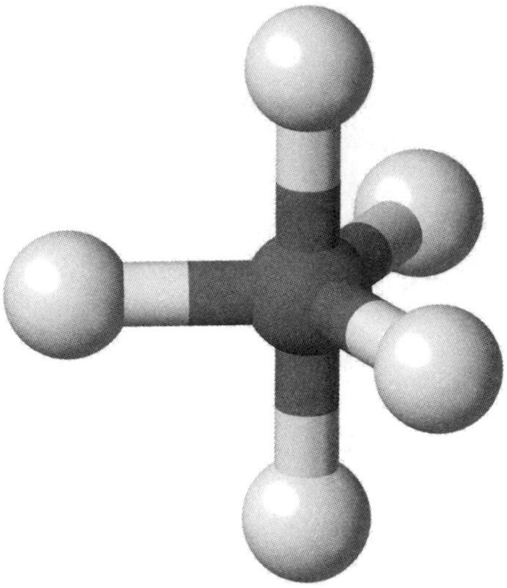
SeO_3	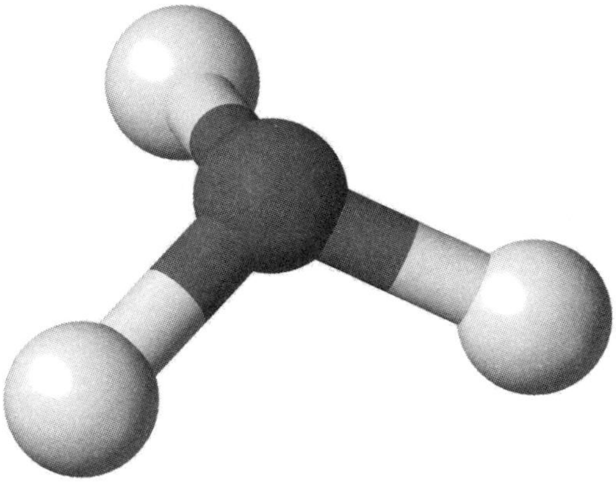
IF_6	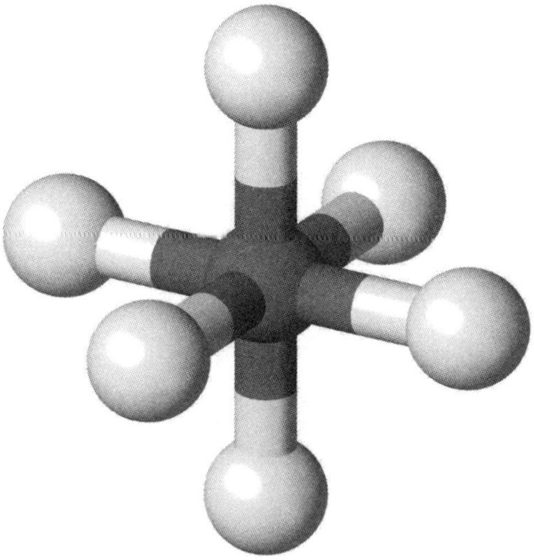

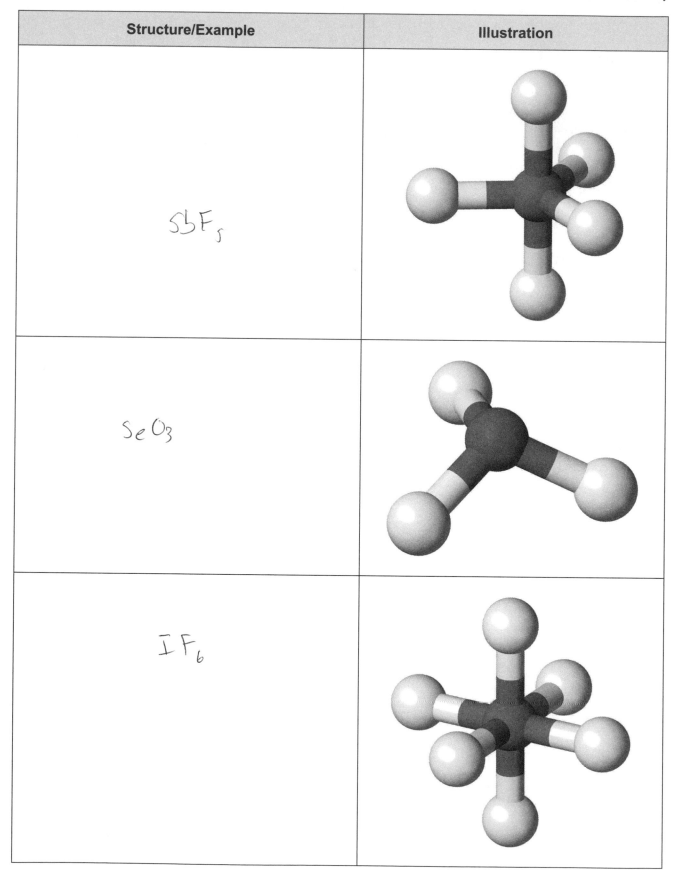

Structure/Example	Illustration
SiF_4	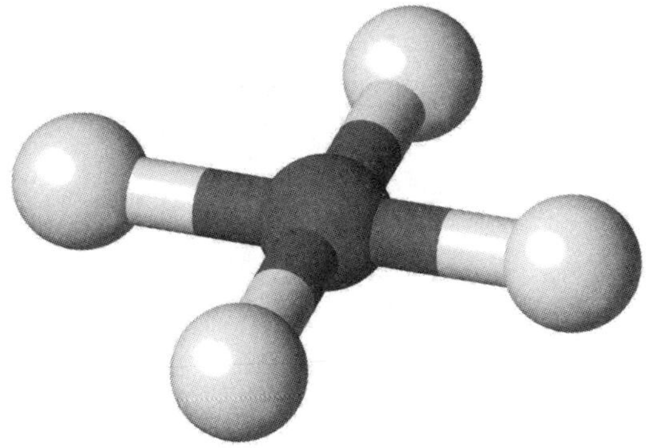
BrF_5	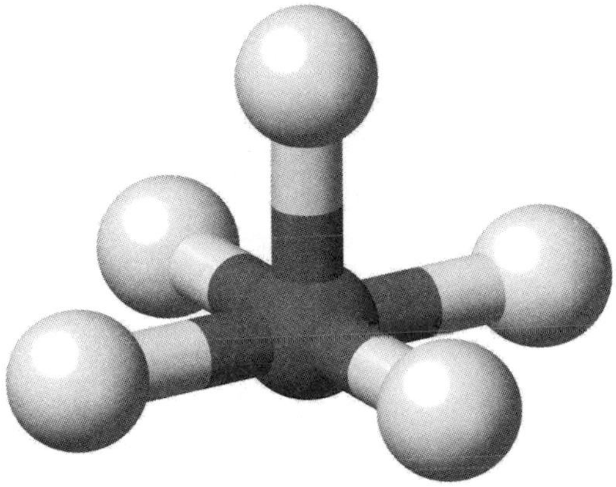
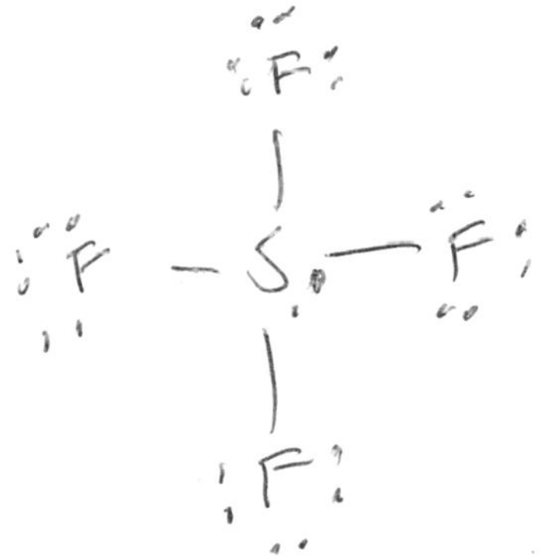	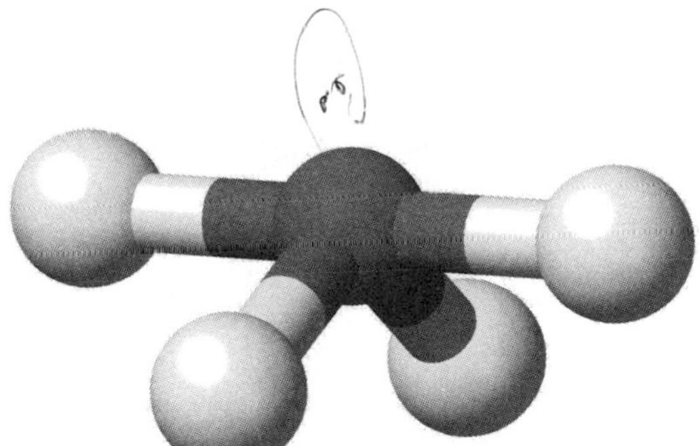
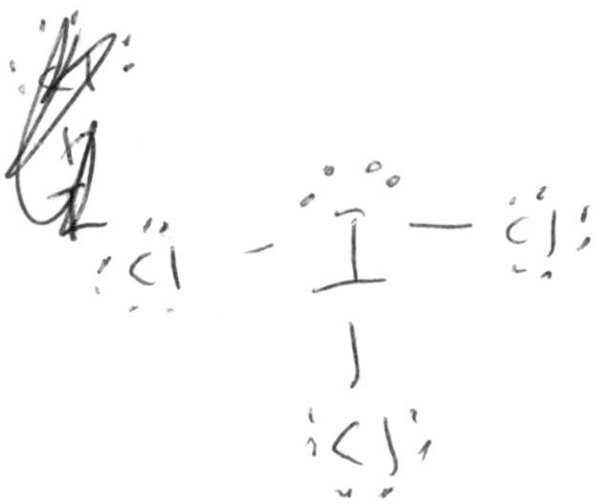	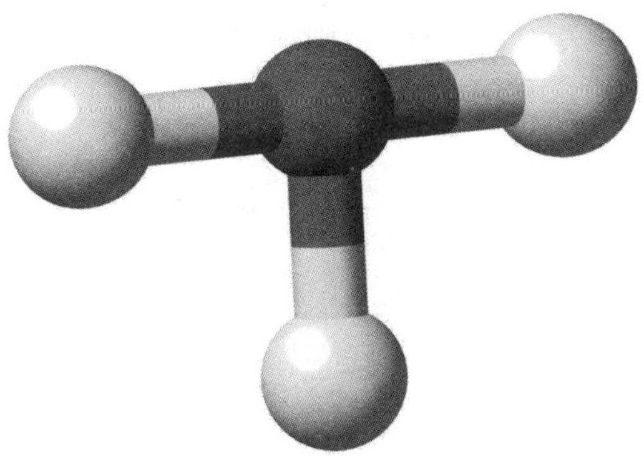

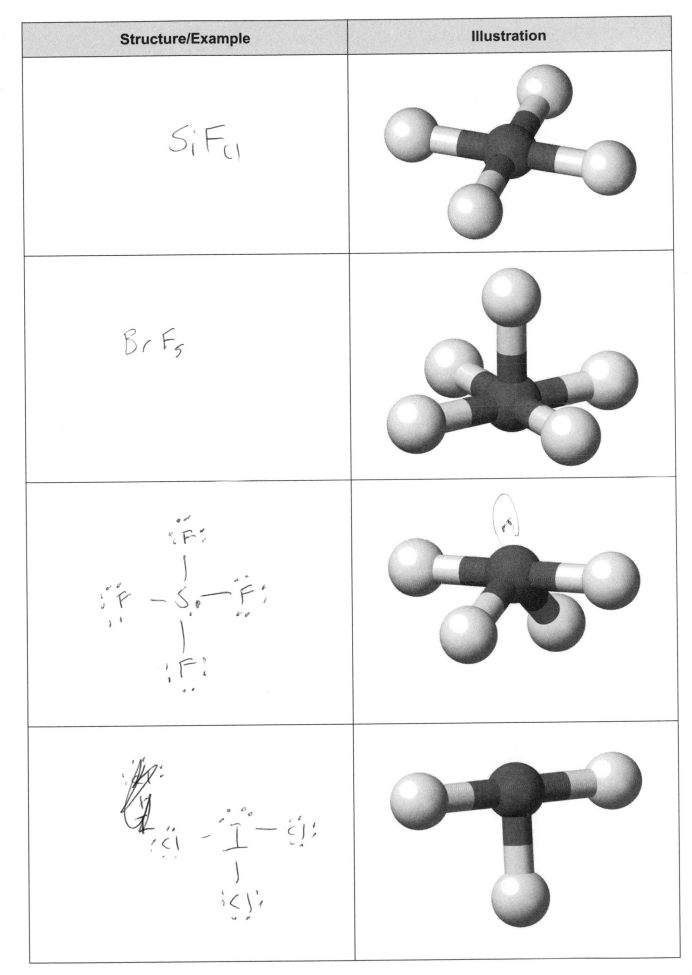

PROBLEMS

1. An article in a journal, *Inorganic Chemistry*, cites both BF_3 and PF_3 as examples of flat or planar molecules with bond angles of 120°. Another article reports the FPF bond angle as 98°. Which report is consistent with the VSEPR model? Explain.

2. Is the shape of OCS like that of CO_2 or SO_2? Identify which are linear and which are bent.

Activity 09-2

Electronegativity and Bond Polarity

Why?

Electronegativity is a measure of the ability of an atom in a molecule to attract electrons. The difference in the electronegativities of two atoms profoundly affects the properties of the chemical bond between the two atoms and, consequently, has dramatic effects on the physical and chemical properties of materials. You therefore need to be able to identify polar bonds and polar molecules from atomic electronegativities and molecular geometries (i.e., shapes).

Learning Objectives

- Master the use of electronegativity to predict bond characteristics
- Identify how bond characteristics affect material properties

Success Criteria

- Correctly identify bonds in order of increasing polarity
- Correctly identify polar molecules

Prerequisites

- **Activity 07-5:** *Periodic Trends in Atomic Properties*
- **Activity 09-1:** *Valence Shell Electron Pair Repulsion Model*

Model: Electronegativity (EN) and Bond Characteristics

ΔEN	Ion/Covalent Character
> 1.7	Mostly ionic
0.4 – 1.7	Polar covalent
< 0.4	Mostly covalent
0	Nonpolar covalent

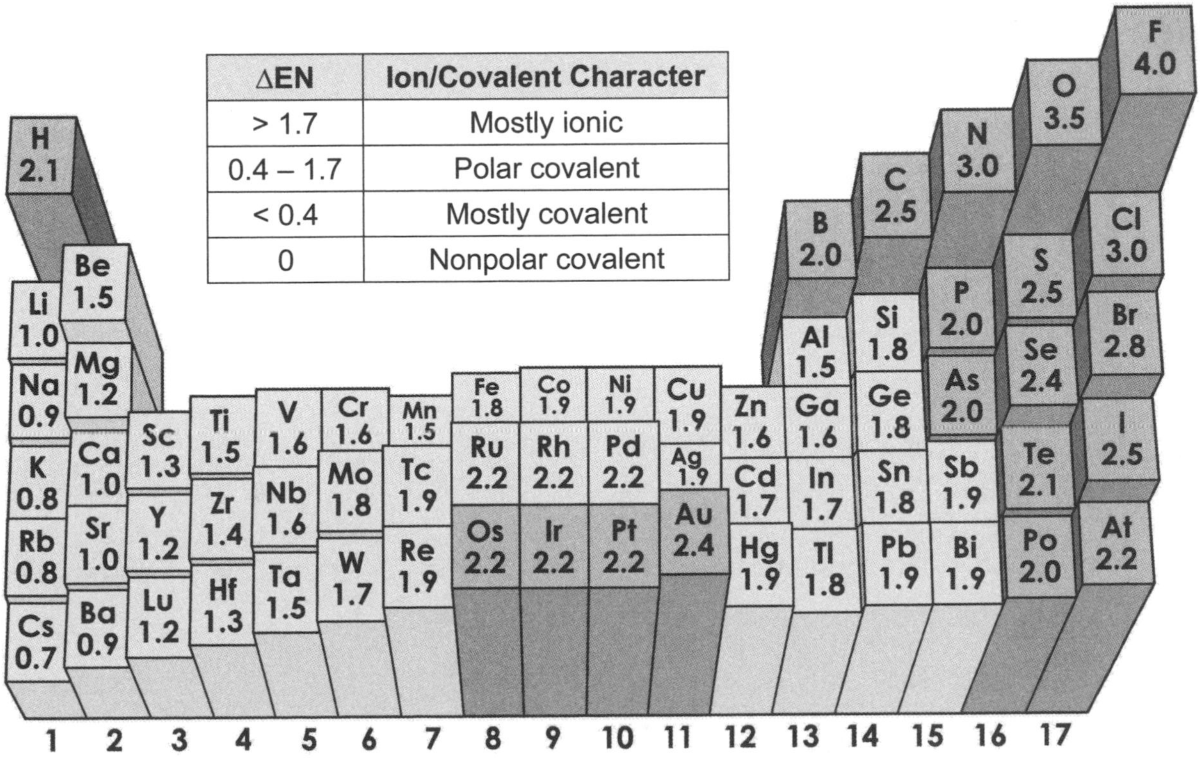

KEY QUESTIONS

1. What is the general trend in electronegativity of atoms across the rows of the Periodic Table?

2. What is the general trend in electronegativity of atoms down the columns in the Periodic Table?

3. What kind of a bond is formed from two atoms that have the same electronegativity?

4. What kind of a bond is formed from two atoms that have very different electronegativities?

5. What is the relationship between electronegativity and bond polarity?

EXERCISES

1. Using position in the Periodic Table as the criterion (do not look at a table or chart of electronegativity values), arrange the elements in each of the following groups in order of increasing electronegativity. Use the < symbol in your arrangement.

(F, C, N, O)	C < N < O < F
(Cl, F, I, Br)	I < Br < Cl < F
(O, S, F)	S < O < F
(Cs, Ca, Cu, S, Se, Cl)	Cs < Ca < Cu < Se < S < Cl
(As, F, Sb, Cl, S, Se)	

2. Perform tasks a) through d) for each of the following groups of bonds:

   (C-F, Si-F)   (C-F, C-Cl, C-Br)   (N-O, N-F, C-F)   (Cl-Cl, F-Cl, F-Br, Br-Cl)

   a) Order the bonds in order of increasing bond polarity. Use the < symbol in your arrangement.

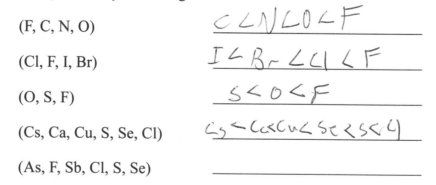

b) Identify the atom that carries a slight positive charge in each bond.

c) Use an arrow (+⟶) to indicate the direction and extent of electron density shift in each bond, based on electronegativity values, with no arrow representing a nonpolar covalent bond. The arrowhead points to the atom that is more electronegative. These arrows are called *bond dipoles*.

d) Classify each bond as mostly ionic, polar covalent, or nonpolar covalent based on electronegativity values.

3. Identify whether or not metals and nonmetals always form mostly ionic bonds. Can you find metal-nonmetal pairs that produce mostly covalent or polar covalent bonds?

INFORMATION

The overall molecular dipole is determined by adding the individual bond dipoles. In this addition process, both the magnitude of the bond dipole and the geometry of the molecule contribute because, if two bond dipoles with the same magnitude are pointing in opposite directions, then they offset each other and the molecule is nonpolar.

EXERCISES

4. Determine which of the following molecules are polar and show the directions of the molecular dipoles. Note that lone pairs are not shown in the structures. Use your textbook to learn the meaning of *cis* and *trans* if you do not already know.

Molecule	Structure	Polar or nonpolar?
CO_2	O=C=O	nonpolar
H_2O	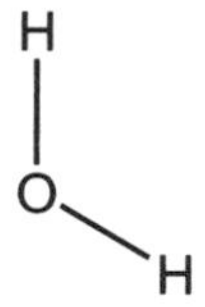	polar
NH_3	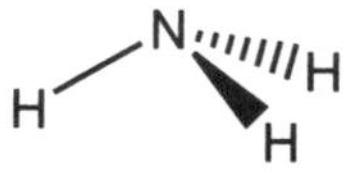	polar
NF_3		~~non~~ polar
NO_3^-	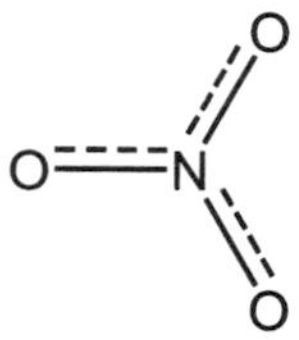	nonpolar
$CHCl_3$	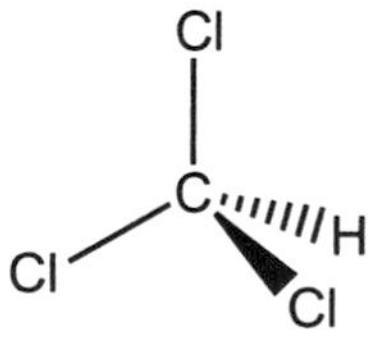	polar
BF_3	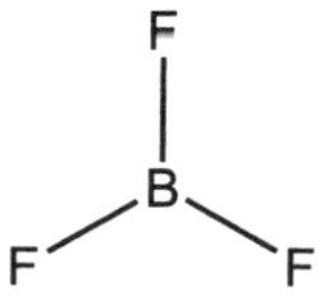	non polar
CH_2F_2	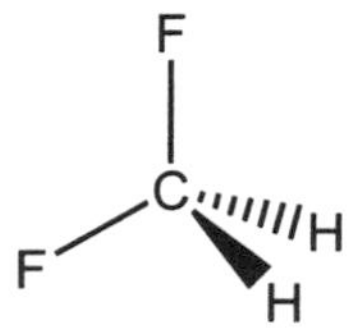	polar

Molecule	Structure	Polar or nonpolar?
cis- PCl_3F_2	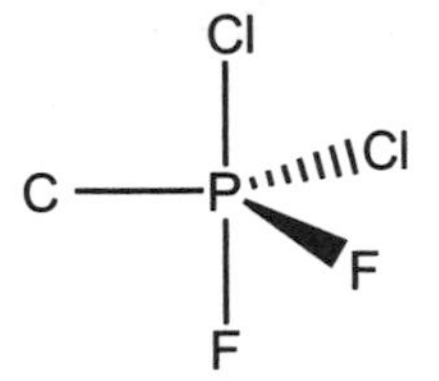	polar
trans- PCl_3F_2	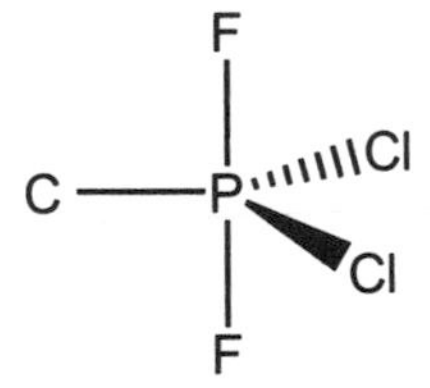	nonpolar
CH_3CH_3	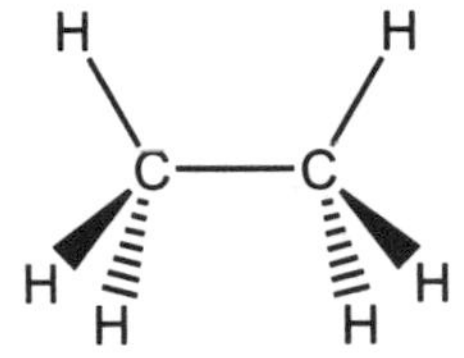	nonpolar
$CH_3CH_2CH_3$	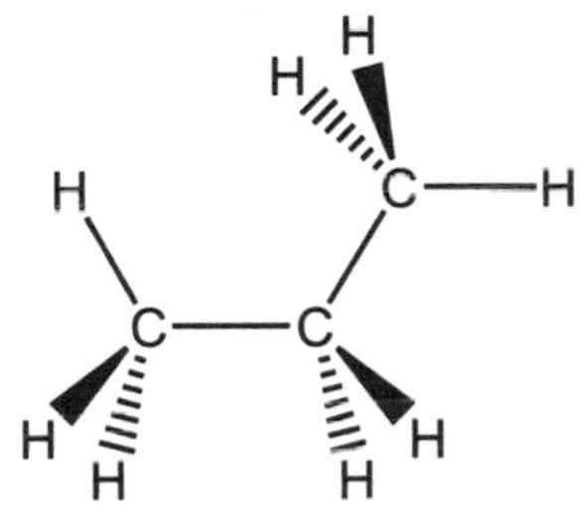	nonpolar
CH_3NH_2	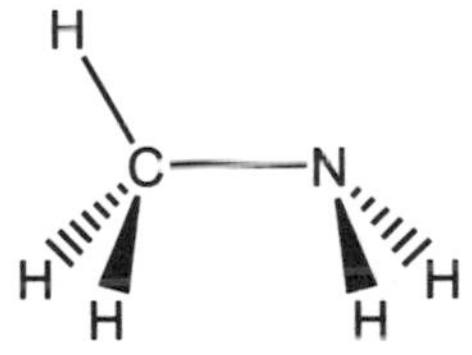	polar
CH_3OH	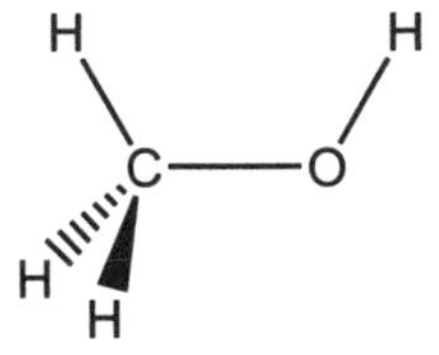	polar
CH_3CH_2OH	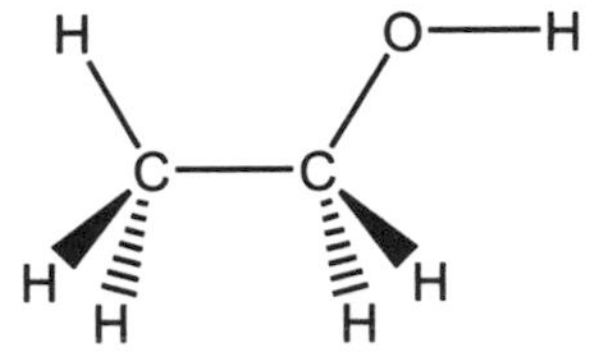	polar

Activity 09-3

Hybridization of Atomic Orbitals

Why?

Molecular orbitals are mathematical functions that describe electrons in molecules in the same way that atomic orbitals describe electrons in atoms. **Activity 08-4:** *Molecular Orbital Theory* illustrated how molecular orbitals can be formed from atomic orbitals. Although molecular orbitals can also be written in terms of other mathematical functions, chemists like to use atomic orbitals because this approach appeals to their intuition about how molecules are produced from atoms and how the electrons are arranged around the atoms in a molecule. Atomic p-orbitals are oriented at angles of 90° to one another, but bond angles in molecules are rarely 90° to one another. To account for bond angles that differ from 90°, atomic orbitals can be combined to form new orbitals called *hybrid orbitals*. The process of combining the atomic orbitals is called *hybridization*. You can use the properties of the different hybrid orbitals to help you recognize molecular geometries and understand the nature of chemical bonds in molecules.

Learning Objective

- Master the geometrical properties of hybrid orbitals

Success Criteria

- Correctly identify hybrid orbitals
- Relate hybrid orbitals to the geometry or shape of molecules

Prerequisites

- **Activity 08-2:** *Lewis Model of Electronic Structure*
- **Activity 08-4:** *Molecular Orbital Theory*
- **Activity 09-1:** *Valence Shell Electron Pair Repulsion Model*

Model: Methane, Ethylene, and Acetylene

Figure 1

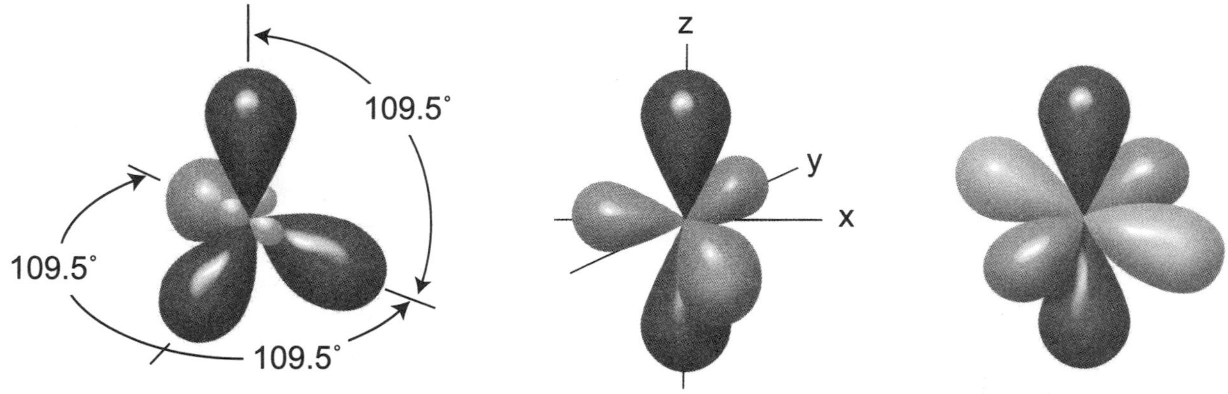

four sp^3 hybrid orbitals

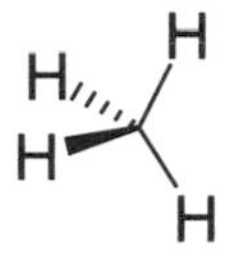

methane

HCH bond angle 109°

three sp^2 hybrid orbitals
one p_z orbital

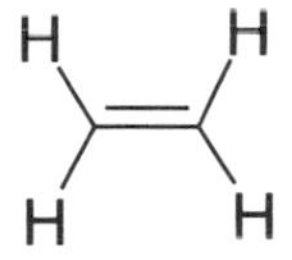

ethylene

HCH bond angle 120°

two sp hybrid orbitals
one p_z orbital
one p_y orbital

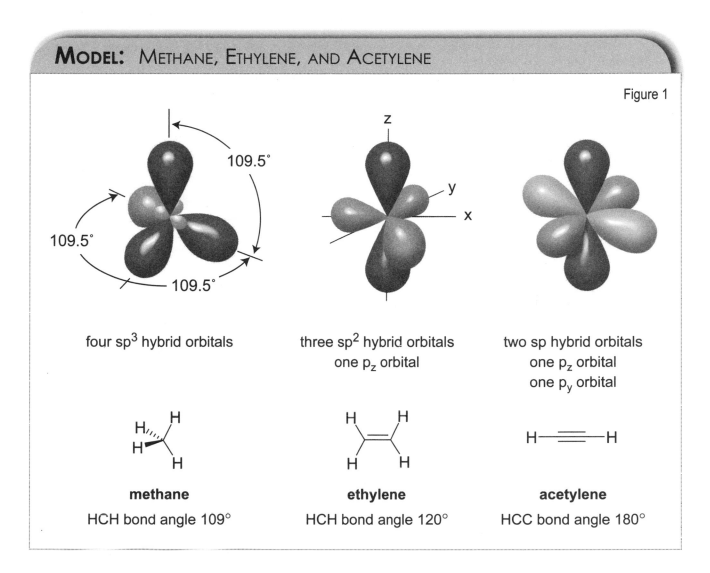

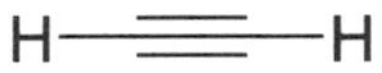

acetylene

HCC bond angle 180°

Tasks

Add the labels sp^3, sp^2, sp, p_y, and p_z with arrows pointing to the orbital to identify the hybrid and atomic orbitals in the model.

Add the angles between the sp^2 and sp orbitals using the sp^3 orbitals in the model as an example.

Key Questions

1. What is the relationship between the number of atomic orbitals used, to the number of hybrid orbitals produced? For example, how many hybrid orbitals does one get by using one s and two p orbitals?

2. Which hybrid orbitals match the bond angles in methane, ethylene, and acetylene? Note that the orbitals represent the locations of the electrons, and the electrons form the bonds, so the orbitals and bonds must line up.

3. In ethylene, which hybrid or atomic orbitals form the sigma-bond (σ) between the carbon atoms, and which form a pi-bond (π)?

4. In acetylene, which hybrid or atomic orbitals form the sigma-bond (σ) between the carbon atoms, and which form the two pi-bonds (π)?

5. Why is it useful to include the concept of hybrid orbitals in the description of molecules?

Exercises

1. Using the Lewis structure and VSEPR model, determine the geometry or shape of each of the following molecules. Identify the hybrid orbitals of the central atom in each, and give the expected bond angles.

Name Molecular Formula	Lewis Structure	Molecular Shape, Expected Bond Angles	Hybridization Around Central Atom
carbon tetrafluoride CF_4	F | F – C – F | F	tetrahedral 109.5°	sp^3
boron trifluoride BF_3 (the octet rule is not followed)	F — B — F | F	trigonal planar	sp^2
beryllium hydride BeH_2 (the octet rule is not followed)	H – Be – H	linear	sp

Name Molecular Formula	Lewis Structure	Molecular Shape, Expected Bond Angles	Hybridization Around Central Atom
sulfur dioxide SO_2		tetrahedral bent	sp^3
ammonia NH_3		tetrahedral trigonal pyramidal	sp^3
formaldehyde H_2CO			sp^2
nitrate anion NO_3^-			
carbonate anion CO_3^{2-}			
ozone O_3			

PROBLEMS

1. Given the following structure, number the carbon atoms 1 through 5 starting from the left.

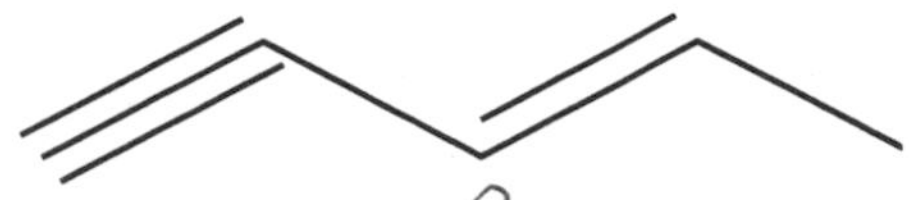

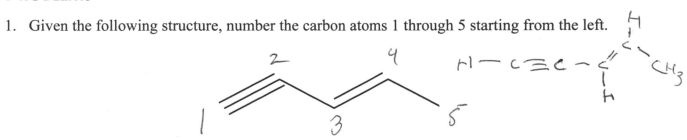

a) Identify the hybrid orbitals on each carbon atom. Be careful; the actual geometry may differ from the way the molecule is drawn. Use the bonding properties of carbon to guide you.

1 ________ **2** ________ **3** ________ **4** ________ **5** ________

b) Determine the local geometry of each carbon atom and predict the bond angles around each carbon atom.

Carbon Atom	Local Geometry	Bond Angles
1		
2		
3		
4		
5		

2. Identify how the atomic orbitals of N, C, and O are hybridized in formamide. The formamide structure is a fundamental linking unit in proteins. Explain your reasoning.

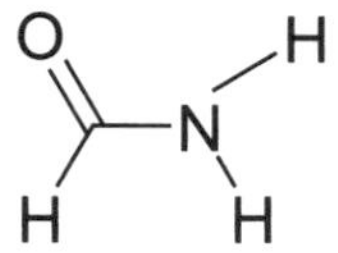

3. If someone told you that formamide was a planar molecule, what would you propose for the hybridization of the N, C, and O atoms?

Activity 10-1

Interactions between Atoms and Molecules

Why?

Strong interactions between atoms, called *covalent bonding*, cause molecules to form. Weaker interactions between atoms and molecules, called *noncovalent interactions*, cause gases to condense and form liquids and solids. The noncovalent interactions are also important in the reactions of enzymes, in the function of nucleic acids, and in determining the structure of proteins. The properties of liquids and solids depend upon the nature and strength of the noncovalent interactions. With an understanding of the types of interactions between atoms and molecules and their relative strengths, you can predict the properties of many materials.

Learning Objectives

- Develop an understanding of covalent and noncovalent interactions between atoms and molecules in terms of their relative magnitudes
- Relate molecular structure to the types of noncovalent interactions
- Relate properties of materials to the noncovalent interactions

Success Criteria

- Identify types of noncovalent interactions from molecular structure
- Predict properties of materials from molecular structure and concomitant noncovalent interactions

Prerequisites

- **Activity 03-1**: *Molecular Representations*
- **Activity 03-2:** *Nomenclature: Naming Compounds*
- **Activity 09-1:** *Valence Shell Electron Pair Repulsion Model*
- **Activity 09-2:** *Electronegativity and Bond Polarity*

Do You Agree or Disagree?

Give your reason for agreeing or disagreeing with each of the following statements.

1. Xenon has a higher boiling point than neon because xenon atoms are heavier than neon atoms.

2. The forces between water molecules are attractive and not repulsive.

MODEL: INTERACTIONS BETWEEN ATOMS AND MOLECULES

One measure of the strength of the interaction between two atoms or molecules is the energy required to separate them. This energy is called the *dissociation energy*. Dissociation energies for some interacting systems are given below.

Figure 1

System		Dissociation Energy kJ/mol
2 Ne		0.29
2 Xe	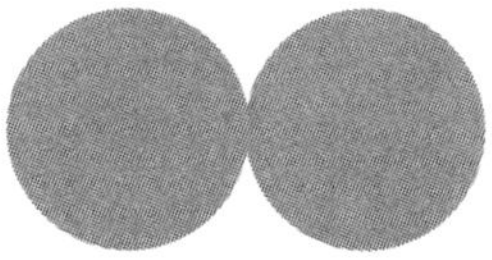	1.84
2 H_2O antiparallel dipoles	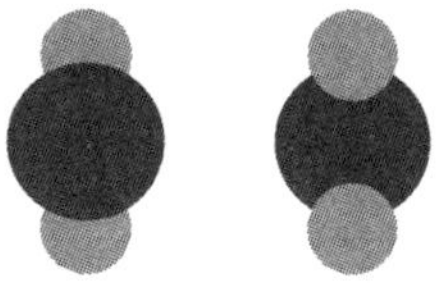	8.20
2 H_2O parallel dipoles		repulsive
2 H_2O head-to-tail		21.9
HO-H		427
NaCl	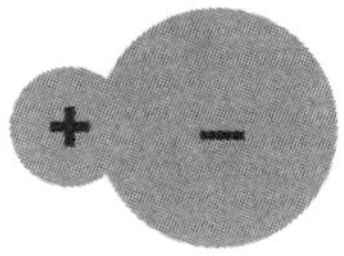	507

Key Questions

1. Which system in the model has a repulsive interaction? Which system has the weakest interaction, in terms of the dissociation energy?

2. Of the systems in the model, which two have the strongest interactions?

3. Which system in the model would you expect to have the lowest boiling point and which would you expect to have the highest? Explain.

4. Why do you think more energy is required to separate two xenon atoms than is required to separate two neon atoms? Remember that electrical forces, which depend on the number and charge of the particles, are stronger than gravitational forces, which depend on the mass of the particles.

5. Why do you think that energy is required to separate two water molecules in the anti-parallel configuration, but in the parallel configuration water molecules repel each other and energy is required to push them together? (Remember that oxygen is more electronegative than hydrogen. This means that the oxygen in water is slightly negatively charged, and the hydrogen is slightly positively charged.)

6. Why do you think that water in the head-to-tail orientation has the highest dissociation energy as compared to the anti-parallel and parallel orientations in the model?

7. What insight has your team obtained so far about why some substances are gases, yet others are liquids or solids at room temperature?

Information

Atoms and molecules interact with each other due to the attraction between opposite charges and the repulsion between like charges. The attractive interactions lead to the formation of molecules, liquids, and solids. When a solid melts or a liquid boils upon being heated, molecules acquire sufficient energy to overcome the attractive noncovalent forces and move away from each other. There is not enough energy, however, to overcome the covalent forces and break the bonds that hold the molecule together.

Interactions between atoms and molecules are given different names depending upon the characteristics or structural features of the atoms and molecules involved. The names are *covalent bonding*, *ionic*, *London dispersion*, *hydrogen bonding*, and *dipole-dipole*.

Covalent bonding interactions between atoms hold molecules together. These bonds are caused by atoms sharing electrons. Covalent bonds are strong, and it requires considerable energy to separate the atoms.

Ionic interactions result when electrons are not shared by atoms, but are transferred from one atom to another. The interaction is between a positively charged cation and a negatively charged anion.

London dispersion interactions occur in all substances. They result from fluctuations in the charge distributions that cause one part of an atom or molecule to be momentarily positively charged and another part to be momentarily negatively charged. These instantaneous fluctuations result in attractive interactions. The noncovalent interactions for atoms and nonpolar molecules result only from the London dispersion interactions caused by these fluctuations.

In *hydrogen bonding*, a hydrogen atom bonded to one of the small, electronegative atoms (oxygen, nitrogen, or fluorine) has a slight positive charge and is attracted to electrons on another of these electronegative atoms.

Dipole-dipole interactions are the interactions between permanent dipoles. Dipoles are formed in molecules because different atoms have different electronegativities. As a result, the electronegative end of the molecule attracts electrons and has a negative charge, and the opposite end has a positive charge. The dipoles in different molecules attract each other when the positive region of one molecule is close to the negative region of another. The dipoles repel each other when like charges are close to each other.

Got It!

1. Oxygen, O_2, and methanol, CH_3OH, have the same mass: 32 amu. Explain why oxygen is a gas, yet methanol is a liquid at room temperature.

2. Examine the reasons that you gave at the beginning of this activity for agreeing or disagreeing with the following statements. Revise and improve your answers by providing as much detail as possible, using the knowledge that you have gained so far from this activity.
   a) Xenon has a higher boiling point than neon because xenon atoms are heavier than neon atoms.

   b) The forces between water molecules are attractive and not repulsive.

Exercises

1. Compare the systems in the table below with those in the Model and identify the dominant interaction as one of the following: London dispersion, covalent bonding, hydrogen bonding, dipole-dipole, or ionic. Enter the name in the table below.

System	Type of Interaction
2 Ne	
2 Xe	
2 H_2O anti-parallel	
2 H_2O parallel	
2 H_2O head-to-tail	
HO-H	

2. The boiling points of butane, acetone, and propanol are 0 °C, 56 °C, and 97 °C, respectively. These molecules, shown below, have about the same number of atoms, electrons, and molar masses, yet their boiling points are quite different. Identify the dominant noncovalent interaction present for each of these molecules and explain, in terms of these interactions, why the boiling points are so different.

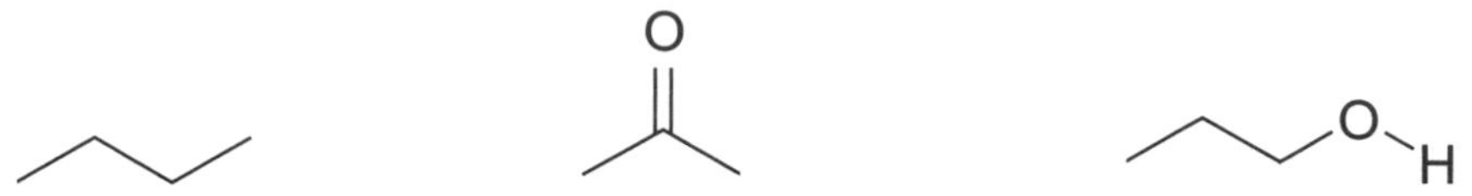

3. Hexane, 2-pentanone, and 1-pentanol are larger than the molecules mentioned in Exercise 2. Their boiling points are 69 °C, 102 °C, and 137 °C, respectively. For each of the types of interactions that you identified in Exercise 2, explain why the boiling point increases with the size of the molecule. Use your observations regarding the interactions between neon atoms and between xenon atoms, and the fact that xenon has more electrons than neon to help you identify the interaction that increases with the number of electrons in the molecule.

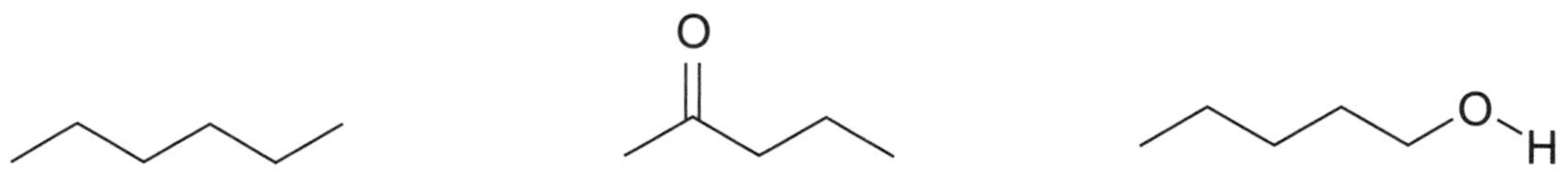

4. Identify the most important type of noncovalent interaction for each of the following substances: $BaSO_4$ (a salt), Ar, NO_2, and HF.

5. Explain why the following compounds have different boiling points: HCl (– 85 °C), HF (20 °C), $TiCl_4$ (nonpolar, 136 °C), and CsCl (a salt, 1290 °C)

6. Identify which of the following does not form hydrogen bonds: HF, H_2O, CH_4, and NH_3. Explain.

Got It!

3. Do you expect that adding an electronegative atom to an alkane will increase or decrease the boiling point? Explain.

4. Do you expect that adding a hydroxyl group to an alkane will increase or decrease the boiling point? Explain.

5. Do you expect that adding alkyl groups to a molecule will increase or decrease the boiling point? Explain

6. Why is the boiling point of ammonia higher than the boiling point of methane?

7. What type of interaction leads to the very high melting point of sodium chloride?

8. What type of interaction leads to the high boiling point of water?

Problems

1. Of the following (a through e), which increase and which decrease, as the strength of the attractive noncovalent interactions increases? Explain.
   a) Deviations from the ideal gas law

   b) Boiling temperature

   c) Sublimation temperature

   d) Vapor pressure

   e) The energy it takes to vaporize 1 mol of the substance

Activity 10-2

Intermolecular Interactions: Water and Organic Molecules

Why?

All living things consist of molecules. In order for these molecules to function as a living system, they must be organized into supramolecular structures such as receptors, DNA, and cells. The organization of the organic molecules of life into living systems involves non-covalent interactions such as hydrogen bonds and dipoles. The interaction of dipolar water with the hydrophobic and hydrophilic parts of organic molecules plays a pivotal role in assembling the molecules of life into living things.

Learning Objectives

- Relate previously learned concepts on non-covalent interactions (dispersion, dipole-dipole, ion-dipole, ion-ion, and hydrogen bonding) to the formation and function of living systems
- An improved understanding of biological structures at the molecular level

Success Criterion

- Ability to explain, in terms of intermolecular interactions, why specific substances are water soluble or water insoluble

Prerequisites

- **Activity 09-1:** *Valence Shell Electron Pair Repulsion Model*
- **Activity 09-2:** *Electronegativity and Bond Polarity*
- **Activity 10-1:** *Interactions between Atoms and Molecules*

Information

Water is an extremely polar substance with a large dipole moment. It also has the ability to form four hydrogen bonds. Salts that dissociate into ions, and molecules that have polar groups (parts) or groups that can hydrogen bond with water, are generally soluble in water. These molecules and groups are called *hydrophilic* or "water-loving." Molecules that only have non-polar groups or groups that cannot form hydrogen bonds with water generally are not soluble in water. These molecules and groups are called *hydrophobic* or "water-fearing." Many molecules of life contain both hydrophilic and hydrophobic groups. These are often called *amphiphilic molecules* or *surfactants*. The solubility of these molecules in water depends upon the balance between the interactions of the hydrophobic and hydrophilic groups with water. If the hydrophobic groups dominate, then the molecule will be insoluble in water. If the hydrophilic groups dominate, then the molecule will be soluble in water. The interaction of water with the hydrophobic and hydrophilic portions of amphiphilic molecules has the amazing ability to organize these molecules into supramolecular structures, structures that are fundamental to life.

Contributed by Frank Fowler and Andisheh Abedini, Stony Brook University

Tasks

The following tasks will guide you in an exploration of the ideas presented in the *Information* section.

1. Using your knowledge of the VSEPR model, draw a three dimensional structure of one water molecule. Illustrate its dipole moment with an arrow and the lone pairs of electrons with dots.

2. Draw an assembly of water molecules that illustrates how one water molecule can form four hydrogen bonds. Represent covalent bonds with a solid line and hydrogen bonds with a dotted line.

3. Draw an assembly of three water molecules forming a six membered ring of alternating oxygen and hydrogen atoms. Use dotted lines for the hydrogen bonds, and solid lines for the covalent bonds.

4. Lithium bromide (LiBr), a salt, is soluble in water. Write the reaction equation for the dissociation of lithium bromide when it dissolves in water. Draw a picture to illustrate the important non-covalent interactions between water and the lithium ion, and water and the bromide ion that are responsible for the solubility of lithium bromide in water. Your picture should reveal the answers to the following two questions: What part of water will be attracted to the lithium cation? What part of water will be attracted to the lithium anion?

5. Ethylene glycol ($HOCH_2CH_2OH$) is very soluble in water. Draw a picture to illustrate the important interactions that occur between water and ethylene glycol to facilitate the solubility of ethylene glycol in water.

6. Octane is not soluble in water. Draw a picture to illustrate the important non-covalent interactions (dipole - induced dipole) between water and octane.

Octane (C_8H_{18}) 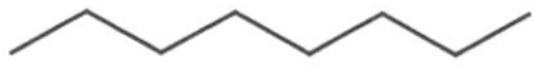

7. Explain why octane is **insoluble** in water, but sodium octanoate is **soluble** in water. Your explanation should include mention of hydrogen bonding and ion-dipole interactions.

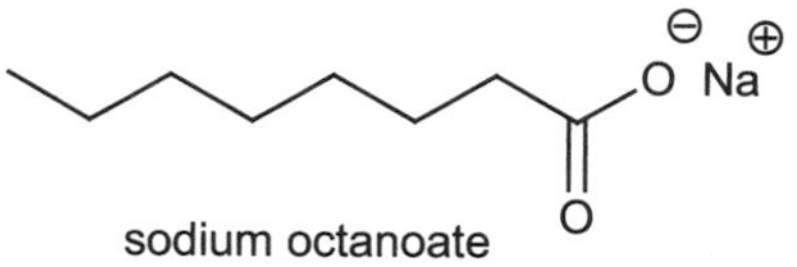

sodium octanoate

8. Explain why acetic acid is soluble in water but stearic acid is insoluble in water.

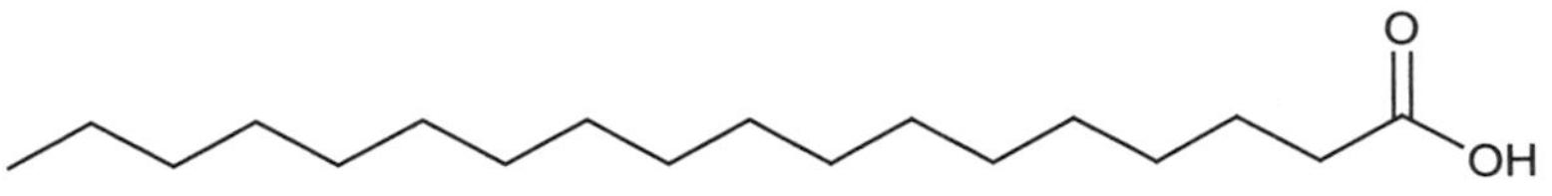

stearic acid

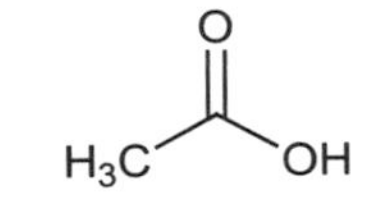

acetic acid

9. A *micelle* is a spherical structure that has the hydrophobic piece of a molecule pointing toward the center, and the hydrophilic piece pointing into the water. Sodium stearate is obtained from stearic acid (see the previous question for the structure of stearic acid) by reaction with sodium hydroxide. Draw a picture of a micelle formed from high concentrations of sodium stearate in water, and describe the intermolecular interactions that lead to the formation of the micelle.

10. Look up the structure of lecithin and, using biological symbolism for lecithin (), sketch how it self-assembles into a bilayer when placed in water.

Activity 10-3

Phase Changes in Pure Substances

Why?

A phase diagram is a convenient way to represent which phase (solid, liquid, or gas) or combination of phases is present at different temperatures and pressures. Knowing the relationship of the phase to the temperature and pressure is necessary in order to determine the suitability of a material for a particular application (e.g., a high-temperature engine component or a cooking oil). Phase diagrams are used, for example, to plan the synthesis of artificial diamonds and new materials or to deduce the history of geological samples. Additionally, knowledge of a material's capacity to absorb energy in one phase or while changing phases is essential to controlling temperature and the stability of systems involving chemical reactions and energy transfer.

Learning Objectives

- Interpret the major features of a phase diagram
- Interpret data presented in a heating or cooling curve to determine the physical and thermochemical properties of a pure substance

Success Criteria

- Identify stable phases, at particular temperatures and pressures, from a phase diagram
- Use a phase diagram to identify the temperatures and pressures at which phase changes occur
- Use a heating curve to determine melting and boiling points, specific heat capacities for each phase, and enthalpies associated with phase changes

Prerequisites

- **Activity 06-1:** *Thermochemistry and Calorimetry*
- **Activity 06-2:** *Internal Energy and Enthalpy*

Information

A *phase diagram* summarizes the relationships between phase, temperature, and pressure. It represents a map of the states of matter for a pure substance.

Boundary lines between pure phases represent the temperatures and pressures at which two phases are in equilibrium. A phase change or transition occurs across these boundary lines.

A *triple point* is the temperature and pressure where three boundary lines come together. At these points, three phases are in equilibrium.

A phase diagram can be produced by observing the phase present under specific conditions of temperature and pressure, as energy is added to or removed from the system. A *heating curve* is a plot of the temperature of the substance as a function of the energy added. A *cooling curve* is a plot of the temperature as a function of the energy removed.

Measurement of the amount of energy transferred in a heating or cooling experiment makes it possible to determine the enthalpies of fusion and vaporization, as well as the specific heat capacities of the different phases.

Suggestions contributed by John Goodwin, Coastal Carolina University

Model 1: General Features of a Phase Diagram

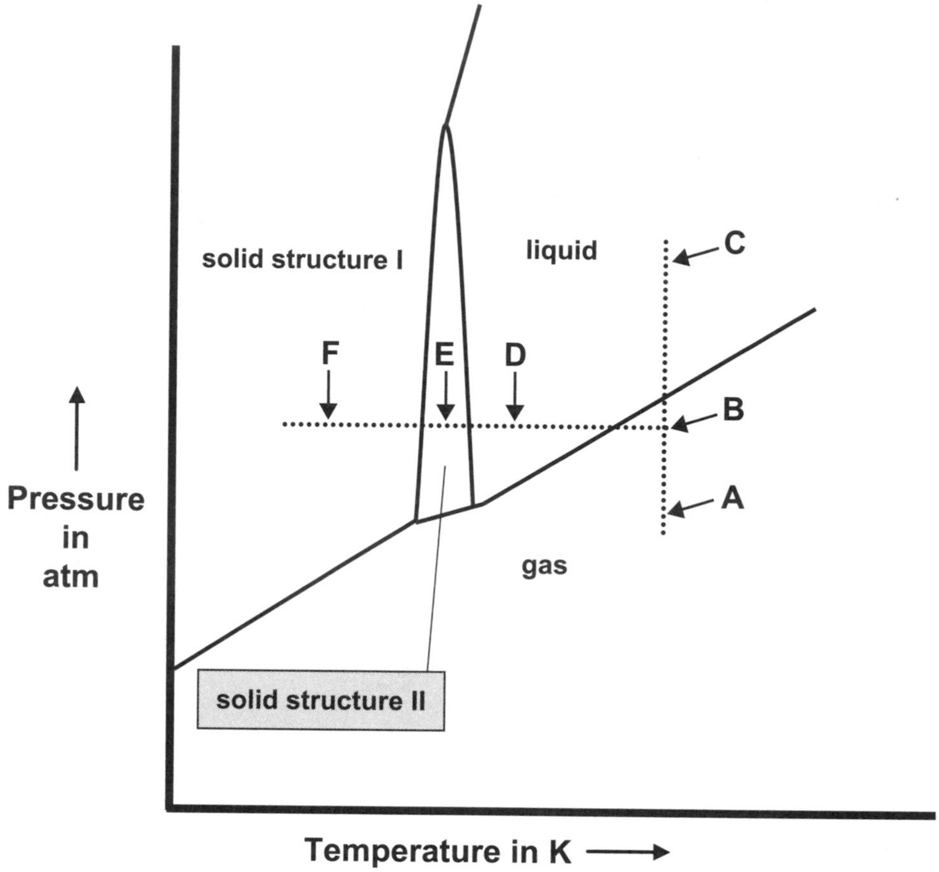

Key Questions

1. What quantities are plotted on the x- and y-axes of a phase diagram?

2. a) How many different phases are shown by the phase diagram in the model?

   b) What are the labels used to identify them?

3. What do the solid lines in a phase diagram represent?

4. a) How many triple points are shown by the phase diagram in the model?

b) What three phases are in equilibrium at each of them?

5. Starting at point A in the model phase diagram, what happens as the pressure increases at constant temperature following the dotted line to point C?

6. Starting at point B in the model phase diagram, what happens as the temperature decreases at constant pressure following the dotted line to point D, then to point E, and finally to point F?

Model 2: General Features of a Heating Curve

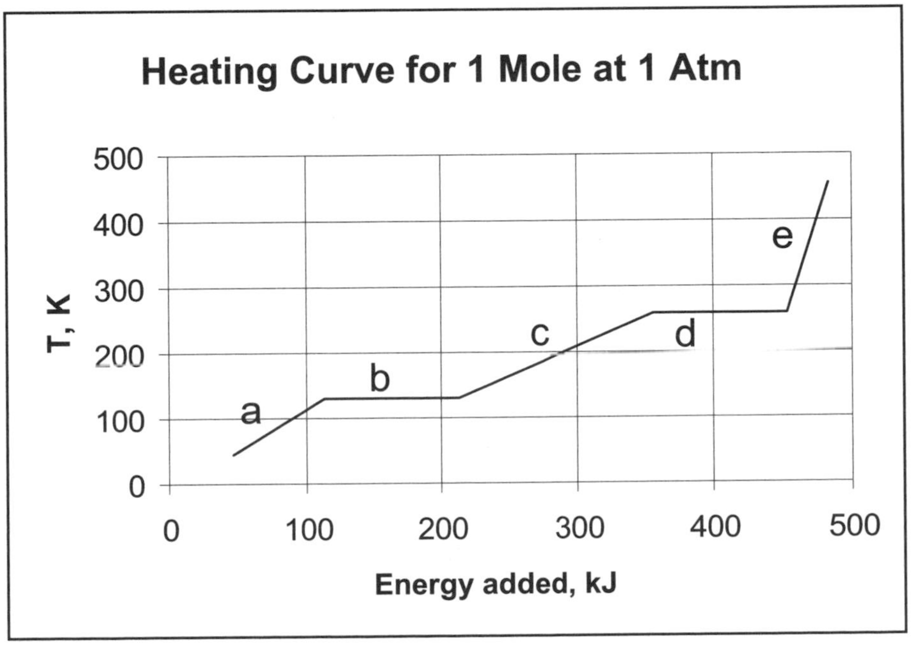

Key Questions

7. What quantities are plotted on the x- and y-axes in a heating curve?

8. Which of the five line segments (labeled a through e) in the heating curve (**Model 2**) correspond to the processes below?

_____ heating the solid _____ heating the liquid _____ heating the gas

_____ phase transition between solid and liquid _____ phase transition between liquid and gas

Exercises

1. The *critical point* in a phase diagram is where the boundary line between the liquid and gas phase ends. Find and label the critical point in the phase diagram in **Model 1**.

2. Sketch a phase diagram for ethanol (CH_3CH_2OH) using the data in the following table. Note that the critical point is the point where the boundary line between the liquid and gas phases ends.

Item	Temperature	Pressure
solid-liquid-gas triple point	–123 °C	4.3×10^{-9} atm
melting point	–114 °C	1.0 atm
boiling point	+ 79 °C	1.0 atm
critical point	+ 241 °C	63 atm

3. Use the heating curve in **Model 2** to estimate both the melting and boiling points of the substance.

4. Use the heating curve in **Model 2** to estimate values for the following parameters for the substance. Assume that the molar mass of the substance is 150 g/mol.

   a) molar heat of fusion

   b) molar heat of vaporization

   c) specific heat capacity of the solid

   d) specific heat capacity of the liquid

   e) specific heat capacity of the gas

Activity 10-4

Unit Cells of Crystalline Solids

Why?

A unit cell is the smallest group of atoms or molecules in a crystal. A unit cell can be used to reproduce the entire crystal by stacking many unit cells together in three dimensions without any spaces between them, just like building blocks. The structures of crystalline solids are described in terms of unit cells, and you will find that the structure is the key to understanding many properties of crystalline solids.

Learning Objectives

- Identify unit cells by their names and characteristics
- Determine the relationships between atomic size, unit cell characteristics, and density of the material

Success Criteria

- Accurate determination of the number of atoms in a particular unit cell
- Ability to quantitatively relate the density of the crystal, the radius of the atoms, and the unit cell composition, structure, and size

Information

You might expect that any number of unit cells could produce crystal structures, but in 1848 the French physicist August Bravais proved that only fourteen distinct unit cells, now called *Bravais lattices*, are needed to generate all the possible three-dimensional crystal lattices. In this activity, three of these fourteen will be considered: primitive cubic, body-centered cubic, and face-centered cubic. A cubic unit cell has sides of equal lengths that intersect at 90° angles. Many metals form solids with these cubic structures. To describe the structure of a metallic crystal, the atoms are represented by hard spheres packed next to and on top of each other. The different ways these hard spheres, or balls, can be stacked in the unit cell produce the different unit cells.

Model

r = metallic radius of an atom **a = length of a side of the cubic unit cell**

Primitive Cubic

The atoms are touching along an edge.

$a = 2r$

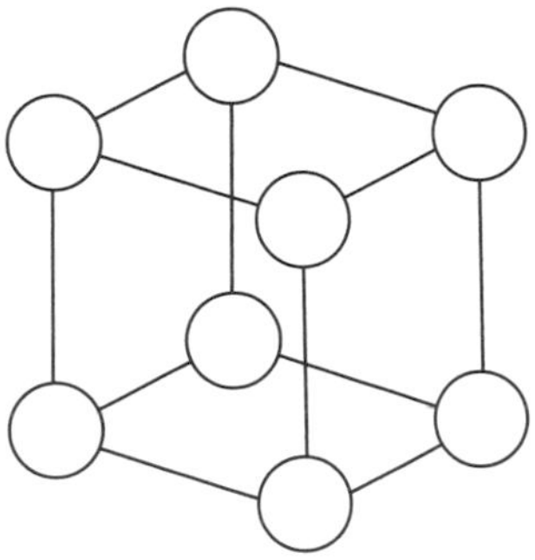
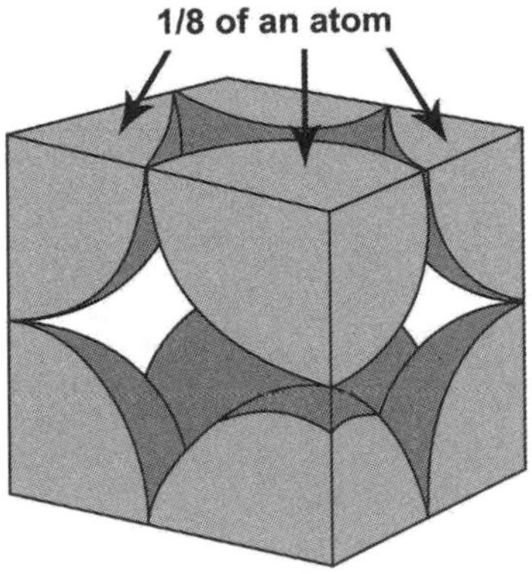

Face-Centered Cubic (aka Cubic Close-Packed)

The atoms are touching along a face diagonal.

$a = \sqrt{8}r$

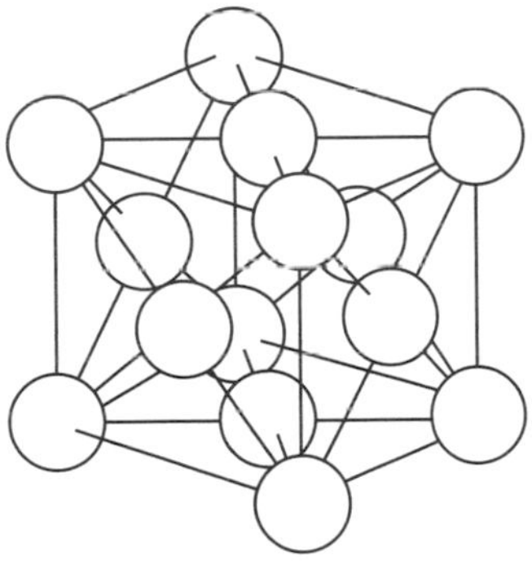
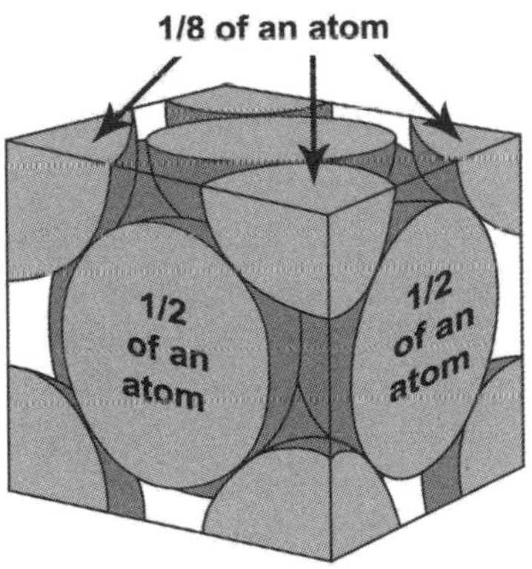

Body-Centered Cubic

The atoms are touching along a body diagonal.

$a = \frac{4}{\sqrt{3}}r$

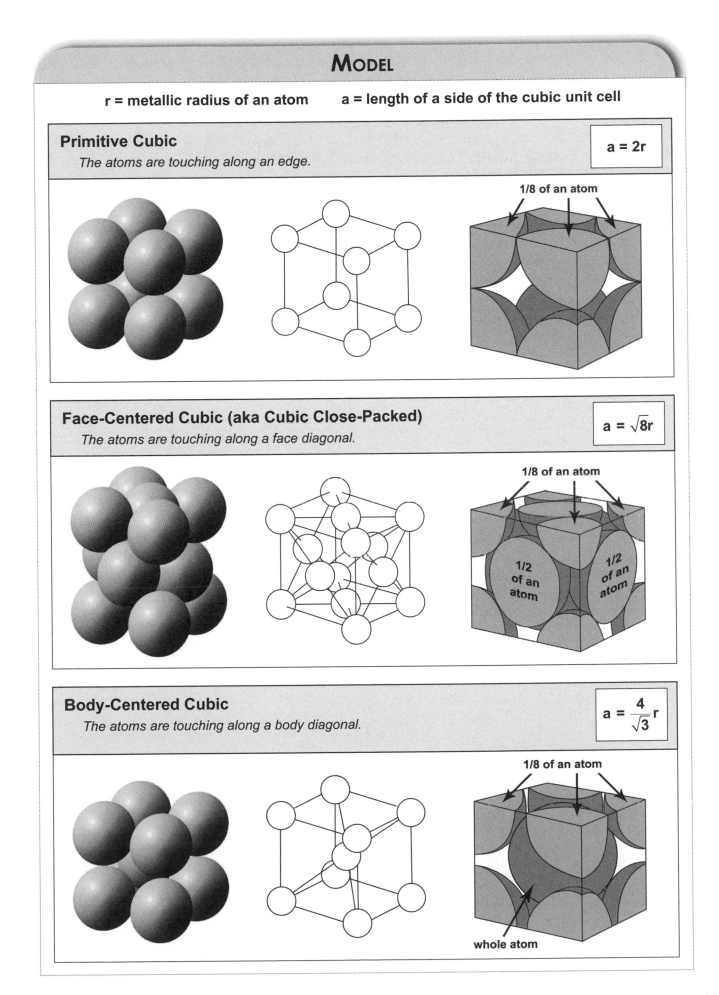

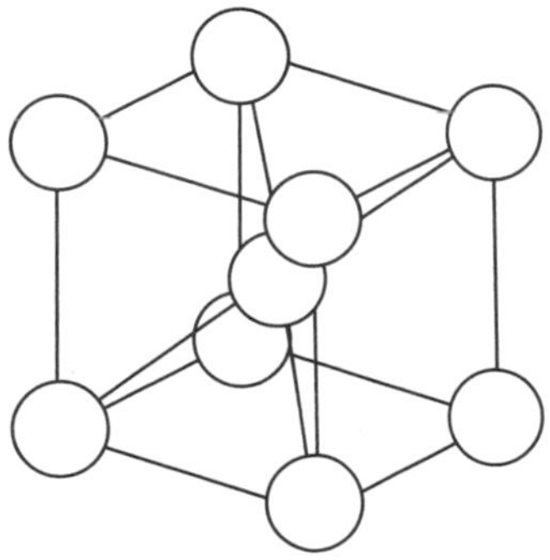
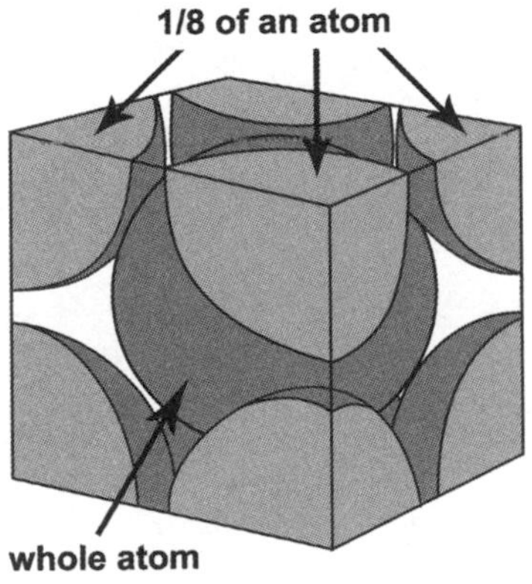

Key Questions

1. What are the names of the three unit cells in the Model?

2. What is the relationship between the metallic radius of the atom and the length of a side for each of the unit cells in the Model?

3. Why is only 1/8 of an atom or 1/2 of an atom associated with each unit cell when that atom is at a corner or the center of a side of the cell, respectively?

4. How many atoms are associated with each of the unit cells in the Model? (Sum the fractional and complete atoms.)

5. For each of the three unit cells in the Model, what is the relationship between the radius of the atom and the volume of the unit cell?

6. How can one calculate the mass of the atoms in the unit cell given the molar mass of the element forming the crystal?

7. In view of your answers to Key Questions 5 and 6, how can the density of the metal be calculated from the characteristics of the unit cell?

EXERCISES

1. How many atoms are there in a unit cell of potassium, which crystallizes in a body-centered cubic structure?

2. Use the Pythagorean theorem, when necessary, to show that the relationships given in the Model between the metallic radius of an atom and the side of a unit cell are correct.

3. Determine the metallic radius of a copper atom from x-ray diffraction data which reveal that the structure is face-centered cubic with a unit cell length of 361.6 pm.

4. Using the x-ray diffraction data given in Exercise 3, determine the density of a copper crystal.

PROBLEMS

1. Copper crystallizes in a face-centered cubic structure. What is the fraction of the unit cell volume that is occupied by copper atoms, assuming the hard sphere model for the atoms with a radius of 127.8 pm? Does this volume depend upon the composition of the crystal, or is it the same for all face-centered cubic structures?

2. Iridium crystallizes in a face-centered cubic structure with a density of 22.6 g/cm^3, and is the densest of all known elements. What is the metallic radius of iridium that is consistent with this information?

3. What is the density of platinum (which crystallizes in a face-centered cubic structure and has a metallic radius of 139 pm)?

ACTIVITY 11-1

The Ideal Gas Law

WHY?

The ideal gas law interrelates the amount of a gas and its pressure, volume, and temperature. It is used to predict or calculate the changes in any of these attributes, due to changes in the others. Understanding and being able to use the ideal gas law is essential whenever you have to work with gases.

LEARNING OBJECTIVES

- Understand the ideal gas law
- Master calculations involving the ideal gas law
- Learn how a molecular formula can be determined from gas density

SUCCESS CRITERIA

- Ability to calculate each of the attributes of a gas (pressure, volume, amount, and temperature), when the others are given or known
- Accurately calculate molar mass from gas density data

PREREQUISITES

- **Activity 03-3:** *Mole and Molar Mass*
- **Activity 03-4:** *Determination of Molecular Formulas*

INFORMATION

The mathematical equation in the Model below is called the *ideal gas law*. It gives an excellent approximation of the relationships among attributes of a gas: pressure (P), volume (V), amount in moles (n), and temperature in Kelvin (T). Studies of gases have shown that the ratio PV/nT has a nearly constant value. This constant is called the Ideal Gas Constant and is symbolized by the letter R. Gases for which (PV) / (nT) = R are called *ideal gases*. Although this law neglects intermolecular interactions and molecular volumes, it can usually be applied as a reasonable approximation to all gases. The ideal gas law is commonly written in the form **PV = nRT**.

Since experiments show that at one atmosphere of pressure and a temperature of 273.15 K (0 °C), one mole of an ideal gas occupies 22.414 L, the value for R is 0.08206 (L atm) / (mol K).

MODEL: THE IDEAL GAS LAW

$$PV = nRT$$

Key Questions

1. How would you state the mathematical equation for the ideal gas law in words rather than using letters as symbols or abbreviations?

2. According to the ideal gas law, if the temperature increases while n and V remain unchanged, what must happen to the pressure?

3. According to the ideal gas law, if the volume increases while n and T remain unchanged, what must happen to the pressure?

4. According to the ideal gas law, if the number of molecules in the container increases while T and P remain unchanged, what must happen to the volume?

5. According to the ideal gas law, if the number of molecules in the container increases while T and V remain unchanged, what must happen to the pressure?

6. At what temperature does an ideal gas exert no pressure?

7. Describe a situation illustrating that it would be nonsense to use a Celsius temperature like 0 °C in the ideal gas law.

Exercises

1. Calculate a value for R and determine the units of the ideal gas constant by using the following data for ammonia:

$$n = 1.000 \text{ mole}, P = 0.1300 \text{ atm}, V = 172.42 \text{ L}, T = 273.15 \text{ K}$$

2. Calculate the volume occupied by 1.00 mole of a gas at 1.00 atm and 273.15 K. Remember this number and note that these values of temperature and pressure are called STP (standard temperature and pressure).

Got It!

1. For the following situations, determine how quantities in the ideal gas law change to keep the PV/nT ratio constant.

   a) If a gas in a sealed jar (V constant) is heated and the Kelvin temperature doubles, how large is the change in the pressure?

   b) If a gas in a jar with a leaky seal (P and V constant) is heated and the Kelvin temperature quadruples, what fraction of the gas escapes?

   c) If the vapor in a cylinder in your automobile engine (T constant) is compressed by the piston to 1/10 of the volume, how large is the change in pressure?

2. According to the ideal gas law, which of the following is/are correct for a gas cylinder of fixed volume filled with one mole of oxygen gas?

   a) When the temperature of the cylinder changes from 15 °C to 30 °C, the pressure inside the cylinder doubles.

   b) When a second mole of oxygen is added to the cylinder, the ratio T/P remains constant.

   c) An identical cylinder filled with the same pressure of hydrogen contains more molecules because hydrogen molecules are smaller than oxygen molecules.

   d) When a second mole of oxygen is added to the cylinder, the ratio T/P decreases by 50%.

   e) None of the above are correct.

PROBLEMS

1. An automobile tire is filled with air at a pressure of 30 lb/in2 at 18 °C. The temperature drops to 0 °C. What is the new tire pressure in lb/in2, assuming the volume of the tire does not change? Some students calculate a pressure of 0, is this correct? If not, what might they have done to obtain a pressure of 0 in their calculation?

2. The valve is opened between a 10 L tank containing a gas at 7 atm and a 15 L tank containing a gas at 10 atm. What is the final pressure in the tanks?

3. Derive an expression for the gas density (m/V) by using the ideal gas law and the relationship between moles of a gas (n) and its mass (m), (M = molar mass).

4. A gaseous sample of a compound has a gas density of 0.977 g/L at 710.0 torr and 100.0 °C. What is the molar mass of this compound? If this compound contains only nitrogen and hydrogen, and is 87.4% nitrogen by mass, what is the molecular formula of the compound?

Activity 11-2

Kinetic Molecular Theory of Gases

Why?

The kinetic molecular theory is a simple model that serves to explain properties of gases, such as the relationships between pressure, volume, temperature, and amount of gas, in terms of the properties of individual gas molecules. Through understanding this theory, you will be able to see how some macroscopic properties of matter are a consequence of the microscopic properties of molecules. You will also be able to explain phenomena like diffusion and effusion, and appreciate why and how real gases differ from ideal gases. (In this discussion of gases, the expression "molecules in a gas" also refers to atoms in atomic gases like He.)

Learning Objective

- Understand and apply the postulates of the kinetic molecular theory

Success Criteria

- Correct analysis of properties of gases in terms of the kinetic molecular theory
- Accurate calculations involving average kinetic energy, root mean square velocity, and diffusion rates

Prerequisite

- **Activity 11-1:** *The Ideal Gas Law*

Model: Postulates of the Kinetic Molecular Theory

1. The volume of individual molecules in a gas is negligible because their size is very much smaller than the distance between them.
2. Gas molecules move constantly and randomly through the volume of the container they occupy at various speeds and in every direction.
3. The forces of attraction and repulsion between molecules in a gas are negligible except when they collide with each other.
4. Gas molecules continually collide with each other and with the walls of the container. The collisions are *elastic*, which means that the kinetic energy of the colliding molecules does not change.
5. The average kinetic energy of the gas molecules, $<KE>$, is proportional to the absolute temperature of the gas. Specifically, using M = molar mass, $<v^2>$ = mean square molecular speed, R = gas constant, and T = Kelvin temperature:

$$<KE> = \frac{1}{2}M<v^2> = \frac{3}{2}RT$$

Key Questions

1. According to the kinetic molecular theory, what assumptions are made regarding the size of the molecules in a gas?

2. According to the kinetic molecular theory, what assumptions are made regarding the interactions between molecules in a gas?

3. According to the kinetic molecular theory, are collisions between molecules visualized more like collisions between sponge balls or collisions between bowling balls? Explain.

4. According to kinetic molecular theory, what information would you need to calculate the mean square molecular speed of a pure substance in the gas phase?

5. How can the postulates of the kinetic molecular theory be used to explain the ideal gas law, i.e., that the pressure of a gas is inversely proportional to the volume of the container (V) and is directly proportional to the number of molecules (n) and the product of temperature and gas constant (RT)?

EXERCISES

1. Make a sketch to illustrate that gas molecules constantly and randomly move at various speeds and in every direction.

2. Calculate the average kinetic energy of 1 mol of hydrogen molecules and of 1 mol of sulfur hexafluoride molecules at 25 °C.

3. Calculate the root mean square speed, $\langle v^2 \rangle^{1/2}$, of an H_2 molecule and of an SF_6 molecule at 25 °C.

4. Compare the ratio of the root mean square speeds of H_2 and SF_6 at 25 °C to the ratio of their molar masses. Are these ratios equal or unequal? Explain.

5. You have a 5.0 L container filled with H_2 gas at standard temperature and pressure.

   a) Use the postulates of the kinetic-molecular theory to predict what happens when the conditions change as described in the table below. Fill in the boxes of the table with *increases*, *decreases*, or *stays the same*. Be able to justify your answers.

Condition	Average Kinetic Energy	Mean Square Velocity	Frequency of Collisions	Force per Collision
temperature increases				
volume increases				
gas added				
H_2 replaced with N_2				

   b) Use your responses in the last row to explain why the pressure of a gas does not depend on the molar mass of the gas.

PROBLEMS

1. Compare the root mean square velocity of a Xe atom at 25 °C with the speed of a jet airliner. Estimate the speed of a jet airliner based on your knowledge and experience.

2. *Diffusion* refers to the spread of one substance through another. The rate of diffusion depends on the speed at which the molecules are moving. Based on the kinetic molecular theory, does the rate of diffusion

   a) increase or decrease with increasing temperature? Explain.

   b) increase or decrease with increasing mass of the molecules? Explain.

3. Does the rate of diffusion change in direct proportion to changes in temperature or mass? Explain.

4. The diffusion rate of a particular gas was measured and found to be 39.9 mm/min. Under the same conditions, the rate of diffusion of SF_6 was 18.7 mm/min. What is the molar mass of the unknown gas? What gas might this be?

5. One group in an introductory chemistry course is arguing that the pressures of an equal number of molecules of hydrogen and sulfur dioxide gases in separate, sealed containers with equal volumes at the same temperature, cannot be equal because a sulfur dioxide molecule has a much larger mass than hydrogen and therefore exerts a much greater force when it hits the walls of the container. They therefore conclude that the pressure of sulfur dioxide in the container should be larger than that of hydrogen. Their intuition seems to be really good. They simply ask, *"Would you rather be hit by a ping-pong ball or a baseball?"* They claim these are real and not ideal gases so the ideal gas law is not relevant. Please write a note to them below, congratulating them if they are right or, if they're not, explaining where they have gone wrong in their logic.

Activity 11-3

Partial Pressure

Why?

The total pressure of a mixture of gases is the sum of the pressures that each component would exert if it were alone in the container. The ideal gas law relates the partial pressure caused by one component in a mixture to the amount of that component present. By using partial pressures to mix gases in the amounts that are needed, you can produce gas mixtures with properties tailored to specific requirements.

Learning Objectives

- Understand the concept of partial pressure
- Master the use of partial pressure and mole fraction in calculations

Success Criterion

- Accurate calculations of partial pressures and mole fractions

Prerequisite

- **Activity 11-1:** *The Ideal Gas Law*

Information

The pressure of a gas in a container is caused by the molecules hitting the walls of the container. If the different gases in a mixture do not interact with each other, as is the case for an ideal gas, then the collisions of one gas with the walls will not depend on the presence of the other gases. The pressure due to each component in the mixture then will be independent of the other components. The total pressure, P_T, then is the sum of the pressures of the different gases in the mixture, P_i.

$$(1)\quad P_T = \sum_i P_i \ \ where\ \ P_i = n_i \frac{RT}{V}$$

In Equation 1, P_i is written in terms of the moles of that gas present, n_i, by using the ideal gas law. The pressure of each component is called the *partial pressure* attributable to that component.

Model: Gases Exert Pressure Independently

Condition	Pressure	Volume	Temperature	Comment
initial	1.0 atm	22.4 L	273 K	1.0 mole of N_2 gas fills the tank.
final	3.0 atm	22.4 L	273 K	2.0 moles of O_2 gas have been added to the 1 mole of N_2 gas in the tank.

Key Questions

1. What was the initial pressure caused by the nitrogen in the tank?

2. What amount of pressure do you think the nitrogen contributed to the final total pressure? Explain your reasoning.

3. If the sum of the pressure due to nitrogen and the pressure due to oxygen must equal the total pressure, what is the oxygen pressure in the Model? Explain your reasoning.

4. What fraction of the pressure is due to nitrogen?

   Note: fraction due to nitrogen = N_2 pressure/total pressure = P_{N_2} / P_T

5. What is the nitrogen mole fraction in the tank?

   Note: mole fraction of nitrogen = moles of N_2 / total moles = n_{N_2} / n_T

6. What is the relationship between the nitrogen pressure ratio in Key Question 4 and the nitrogen mole fraction in Key Question 5? Express this relationship in the form of an equation.

Information

The pressure due to one component of a gas mixture is called the *partial pressure* due to that component. Your response to Key Question 6 is one way of stating *Dalton's Law of Partial Pressures*.

$$\frac{P_i}{P_T} = \frac{n_i}{n_T}$$

Exercises

1. The mole fraction of oxygen in air is 0.209. Assuming nitrogen is the only other constituent, what are the partial pressures of oxygen and nitrogen when the barometer reads 755 torr?

2. Calculate the partial pressure of 2.25 moles of oxygen that have been mixed (no reaction) with 1.75 moles of nitrogen in a 15.0 L container at 298 K.

Making Connections by Solving Problems

(These problems require that you use what you have learned about the ideal gas law and stoichiometry with Dalton's Law of Partial Pressures.)

1. A 25.0 L gas cylinder at 32 °C is filled with 2.33 moles of carbon dioxide and 4.66 moles of nitrogen. Calculate the total pressure and the mole fractions and partial pressures of carbon dioxide and nitrogen in the cylinder.

2. Mixtures of helium and oxygen are used in scuba diving tanks to help prevent "the bends," a condition caused by nitrogen bubbles forming in the bloodstream. If 95 L of oxygen and 25 L of helium at STP are pumped into a scuba tank with a volume of 8.0 L, what is the partial pressure of each gas in the tank? What is the total pressure in the tank at 25 °C?

3. Oxygen gas generated by the decomposition of potassium chlorate is collected over water at 25 °C in a 1.00 L flask at a total pressure of 755.3 torr. The vapor pressure of water at 25 °C is 23.8 torr. How many moles of $KClO_3$ were consumed in the reaction?

$$\mathbf{2\ KClO_3(s) \longrightarrow 2KCl(s) + 3O_2(g)}$$

Activity 12-1

Solutions

Why?

A solution is a homogeneous combination of two or more substances. The major component is the solvent; the other components are the solutes. Knowledge of the measures of concentration, the factors that determine solubility, and the unique properties of solutions enables scientists, engineers, medical professionals, and you too, to predict properties of unfamiliar solutions.

Learning Objectives

- Identify factors affecting solubility
- Define measures of solution composition
- Learn how a solute affects vapor pressure, boiling point, freezing point, and osmotic pressure

Success Criteria

- Predict solubilities for various substances and situations
- Calculate solution concentrations using different units
- Correctly predict changes in solvent properties caused by a solute

Prerequisites

- **Activity 05-2:** *Solution Concentration and Dilution*
- **Activity 10-1:** *Interactions Between Atoms and Molecules*
- **Activity 10-2:** *Intermolecular Interactions: Water and Organic Molecules*

Tasks

Use your textbook as a reference as you perform the following tasks:

1. Complete Table 1 on factors affecting solubility.
2. Complete Table 2 on measures of solution concentration.
3. Complete Table 3 describing the effects of a solute on solvent properties.

Table 1 Factors Affecting Solubility

Note: *In Table 1, identify general trends. Exceptions do sometimes occur; we will learn about those and the reasons they occur later.*

Condition	Effect on Solubility
polar solute/polar solvent	*The solute dissolves.*
nonpolar solute/nonpolar solvent	
polar solute/nonpolar solvent	
nonpolar solute/polar solvent	
increasing the pressure of a gas over a liquid solvent	
increasing the temperature when dissolving a solid	
increasing the temperature when dissolving a gas	

Table 2 Measures of Solution Concentration

Quantity	Formula	Example
mass percent	$= \frac{\text{solute mass}}{\text{total soln mass}} \times 100\%$	20 g of NaCl in 90 g water = (20 g NaCl/110 g soln) × 100% = 18% by mass
mole fraction		0.25 mol NaCl in 5.50 moles water = 0.25 mol NaCl/5.75 total moles = 0.043 mole fraction
molarity (M)		0.25 mol NaCl in 0.50 L of solution = 0.25 mol NaCl/0.5 liter soln = 0.50 M
molality (*m*)		0.25 moles NaCl in 1.5 kg of water = 0.25 mol NaCl/1.5 kg water = 0.17 *m*

In Table 3, identify the qualitative effect that the addition of a solute has on the solvent property.

Table 3 Solute Effects

Property	Qualitative Effect of Solute
vapor pressure	reduces vapor pressure
boiling point	
freezing point	
osmotic pressure	increases osmotic pressure

Key Questions

1. What general rule can you identify in Table 1, regarding the solubility of polar and nonpolar solutes in polar and nonpolar solvents?

2. What is the difference between solutions of sugar and water that are 1 M and 1 *m* according to Table 2?

3. Why do people put salt on ice in the winter time?

4. Is there a correlation between the effect of a solute on the vapor pressure and the boiling point of a solvent? Explain.

EXERCISES

1. Which solvent, water (H_2O) or hexane (C_6H_{14}), would you use to dissolve the following solutes? Explain.

   $(NH_4)_2SO_4$

   HF

   octane (C_8H_{18})

2. A solution is prepared by dissolving 25.00 g of acetic acid (CH_3COOH) in 750.0 g of water. The density of the resulting solution is 1.105 g/mL.

   a) What is the mass percent of acetic acid in the solution?

   b) What is the molarity of the solution?

c) What is the molality of the solution?

d) What is the mole fraction of acetic acid in the solution?

Activity 12-2

Colligative Properties

Why?

Properties of a liquid that depend upon the relative number, and not the identity, of other molecules dissolved in it are called *colligative properties*. You should understand the quantitative as well as the qualitative nature of these properties, as they are important in everyday life as well as in scientific research. For example, salt is thrown on ice to melt it, salty water is used to boil eggs in Denver, and osmotic pressure causes the flow of water through plants, even to the very tops of trees.

Learning Objectives

- Identify the basic colligative properties
- Determine the effects of solutes on boiling point, freezing point, and osmotic pressure

Success Criteria

- Accurate calculations of freezing-point depression, boiling point elevation, and osmotic pressure
- Determination of the molar mass of a solute based upon its effect on the colligative properties of a solvent

Prerequisite

- **Activity 12-1:** *Solutions*

Information

Colligative properties include the lowering of the vapor pressure, the depression of the freezing point, the elevation of the boiling point, and the osmotic pressure caused by adding a nonvolatile solute to a solvent. These properties are all based on the idea that, because a solution consists of fewer solvent molecules and more nonvolatile solute particles, fewer solvent molecules will reach the surface and escape.

Since the volume of a solution changes with changes in temperature, the molarity of the solution will change as well. Molality is therefore the concentration expression used, because it involves mass instead of volume, and is consequently unaffected by temperature.

MODEL: NOTATION AND EQUATIONS FOR COLLIGATIVE PROPERTIES

ΔT_b = elevation in the boiling point

K_b = molal boiling point elevation constant

c_m = molality of the solution

ΔT_f = depression of the freezing point

K_f = molal freezing-point depression constant

Π = osmotic pressure

R = the ideal gas constant

T = temperature in K

RT = a measure of the average translational energy of a collection of molecules

M = molarity of the solution

i = van 't Hoff factor, which is the ratio of the moles of particles in solution to the moles of solute dissolved

Units of pressure 1 atmosphere = 760 torr

The boiling point is elevated: $\Delta T_b = iK_bc_m$

Freezing point is depressed: $\Delta T_f = iK_fc_m$

Solvent will move from the low solute molarity region to the high solute molarity region, even through a separating membrane, producing an osmotic pressure: $\Pi = iMRT$

KEY QUESTIONS

1. What colligative properties are included in this activity?

2. Why is molality rather than molarity used as the concentration in the equations for freezing-point depression and boiling point elevation?

3. How can you determine the molar mass of a polymer by measuring the osmotic pressure produced by dissolving some amount of the polymer in a solvent?

EXERCISES

1. Explain why the freezing-point depression of the following compounds in water (each 1.00 m solutions) increases in the indicated order.

$$\text{glucose} < HOCl < NaCl < MgCl_2$$

2. Determine the boiling point of 1 L of water (K_b = 0.51 °C/m) when 1 oz of salt (28 g sodium chloride) is added to it. Does it make sense for people in Denver to add salt to increase the boiling point of water in this mile-high city? Explain.

3. Determine the freezing point of 1 L of water (K_f = 1.86 °C/m) when 1 oz of salt (28 g sodium chloride) is added to it.

4. Arrange the following aqueous solutions in order of increasing osmotic pressure:

   0.10 M KCl, 0.01 M sucrose, 0.30 M K_2SO_4, 1.0 M HCl, and 0.01 M $CaCl_2$.

5. Calculate the average molar mass of polyethylene when 4.40 g of the polymer is dissolved in benzene to produce 200.0 mL of solution, and the osmotic pressure is found to be 7.60 torr at 25 °C.

6. Sulfur exists in many forms with the general molecular formula S_n. If 0.48 g of sulfur are added to 200 g of carbon tetrachloride, and the freezing point of the carbon tetrachloride ($K_f = 30$ °C/m) is depressed by 0.28 °C, what is the molar mass and molecular formula of the sulfur?

7. Beaker A contains 100.0 mL 1.0 M salt solution (NaCl). Beaker B contains 100.0 mL pure water. Both beakers are placed inside a large container, which is sealed.

   a) Draw a graph showing how both volumes (solution in beaker A and liquid in beaker B) change with time. Explain.

b) Draw a graph showing how both concentrations (the solution in beaker A and the liquid in beaker B) change with time. Explain.

Activity 13-1

Rates of Chemical Reactions

Why?

Chemical kinetics is the part of chemistry that looks at the speed at which reactants are converted into products. An experimental rate law can be determined from such measurements, and this rate law can be used to decide how the reaction occurs. Knowledge of the reaction mechanism and the factors that affect the rate of a reaction makes it possible for the chemist to plan the efficient and cost-effective production of industrial, pharmaceutical, and consumer chemicals. It is essential to understand reaction rates in order to understand how reactions occur.

Learning Objective

- Understand reaction rates and rate laws

Success Criteria

- Use kinetic data to identify a reaction as zero, first order, or second order
- Determine the rate constant for a reaction from kinetic data
- Graph the concentrations of reactants and products as the reaction progresses

Prerequisites

- **Activity 04-1:** *Balanced Chemical Reaction Equations*
- **Activity 05-2:** *Solution Concentration and Dilution*
- **Activity 12-1:** *Solutions*

Information

The *average rate* of a reaction is given by the change in the concentration of a reactant, Δc_r, or a product, Δc_p over some time interval (Δt), provided that the stoichiometric coefficient for the product and reactant is 1. If the stoichiometric coefficient for some species is not 1, then the rate is determined by dividing the concentration of that species by its stoichiometric coefficient.

$$\text{(1)} \quad \text{reaction rate} = -\frac{\Delta c_r}{\Delta t} = \frac{\Delta c_p}{\Delta t}$$

Since the reaction rate is always given as a positive number and the reactant concentration is decreasing, $\frac{\Delta c_r}{\Delta t}$ has been multiplied by –1 in Equation 1.

The *instantaneous rate* of a reaction at a particular time is the limit of the average rate at that time as Δt approaches 0. If the time interval is small, the average rate is a good approximation to the instantaneous rate.

The *initial rate* of a reaction is the instantaneous rate at the beginning of the reaction, when the reactants are mixed together.

Model 1: Kinetics of Ozone Decomposition

Under certain conditions, ozone in the atmosphere decomposes by dissociation:

$$(O_3 \rightarrow O_2 + O)$$

The concentration of ozone as a function of time is given by the graph and table of data below. Note that the curved line in the graph refers to the left scale, which gives the ozone concentration. The straight line refers to the right scale, which gives the natural logarithm of the ozone concentration divided by the initial ozone concentration.

Figure 1

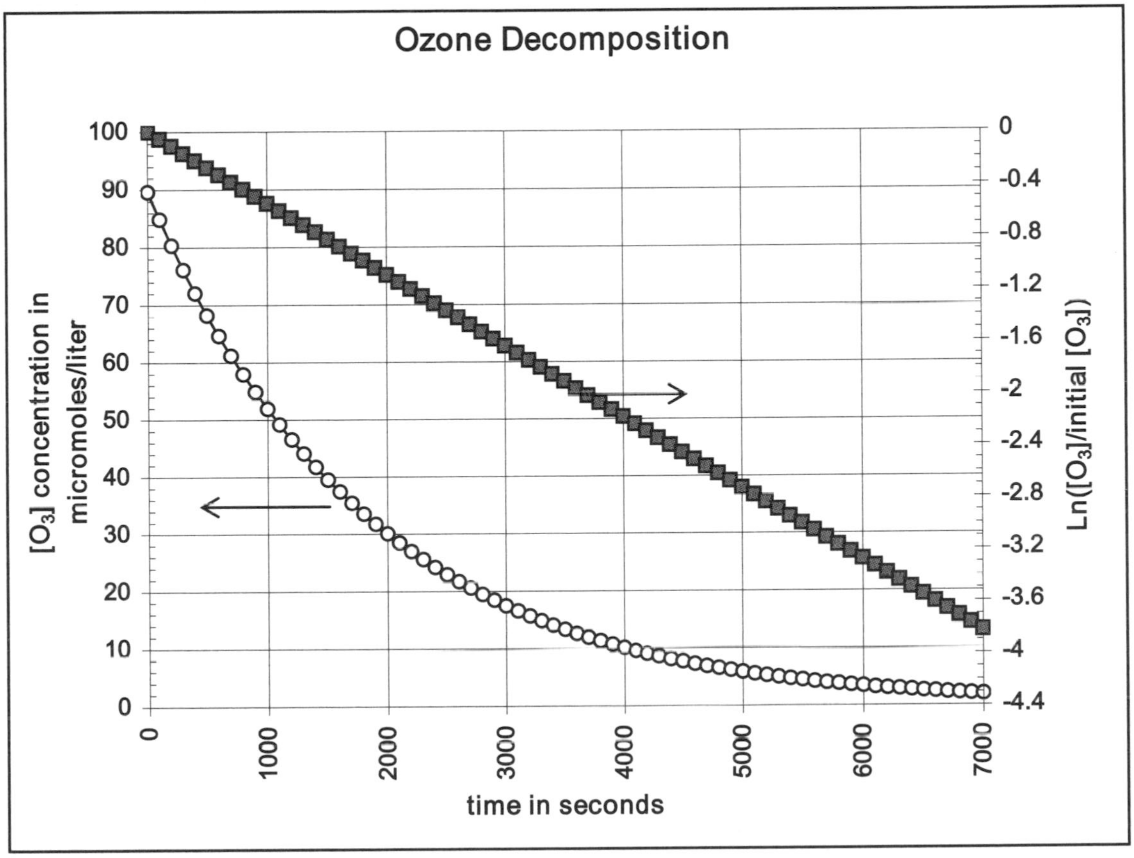

Data on Ozone Decomposition

Time (s)	0	100	900	1000	6900	7000
concentration (10^{-6} M)	89.63	84.87	54.84	51.92	2.07	1.96
$\ln([O_3]/[O_3]_o)$	0.000	– 0.0546	– 0.4914	– 0.5460	– 3.767	– 3.822

Key Questions

1. What quantities are plotted in the graph shown in **Model 1**?

2. What is the relationship between the data in the table and the graph?

3. In the model, how long does it take for half of the ozone to decompose? What is the concentration of ozone after that time? This time is designated as the half-lifetime or the *half-life* of ozone.

4. What is the initial reaction rate of the ozone decomposition reaction? Provide both the magnitude and the units.

5. What is the rate of the reaction after 6900 seconds have passed?

6. What happens to the reaction rate as the concentration of ozone decreases?

INFORMATION

A rate law for a reaction indicates how the rate of the reaction depends on the concentrations of the chemical species involved in the reaction. Generally, rate laws have the following form: the rate of the reaction equals some constant multiplied by a product of concentrations [A], [B], and [C], with exponents x, y, and z, respectively. In equation form:

$$\text{rate} = k\,[A]^x\,[B]^y\,[C]^z$$

An exponent gives the *order* of the reaction with respect to that chemical species. For example, if x and z equal 1, then the reaction is first order with respect to A and C; and if $y = 2$, then the reaction is second order with respect to B. The *overall order* of the reaction is the sum of the exponents in the rate law, which, in this example, would be: $1 + 2 + 1 = 4$.

The order of a reaction often differs from the stoichiometric coefficients, which relate the numbers of reactant and product molecules involved in the reaction. The order of a reaction depends on the reaction mechanism, not on the number of molecules involved in the overall reaction.

Model 2: Rate Laws for Chemical Reactions

In the equations below, the reactant is represented by A; $[A]_t$ represents the concentration of A at time t; $[A]_0$ represents the initial concentration of A; k is the *rate constant*; and *ln* is the natural logarithm. The rate constant is the constant of proportionality in the rate law. The *integrated rate law* is obtained from the rate law by integration of the differential equation.

Zero-order reaction

$A \longrightarrow$ products

rate law $\quad \text{rate} = -\dfrac{\Delta[A]}{\Delta t} = k[A]^0$

integrated rate law $\quad [A]_t = -kt + [A]_0$

First-order reaction

$A \longrightarrow$ products

rate law $\quad \text{rate} - -\dfrac{\Delta[A]}{\Delta t} = k\,[A]$

integrated rate law, exponential form $\quad [A]_t = [A]_0\, e^{-kt}$

integrated rate law, logarithmic form $\quad \ln([A]_t) = -kt + \ln[A]_0$

Or, equivalently, $\ln\left(\dfrac{[A]_t}{[A]_0}\right) = -kt$

Second-order reaction

$A \longrightarrow$ products

rate law $\quad \text{rate} = -\dfrac{\Delta[A]}{\Delta t} = k\,[A]^2$

integrated rate law $\quad \dfrac{1}{[A]_t} = kt + \dfrac{1}{[A]_0}$

Key Questions

7. What are four similarities between zero, first, and second order reactions as represented by **Model 2**?

8. What are three differences between zero, first, and second order reactions as represented by **Model 2**?

9. If the rate constant k increases when the temperature increases, what happens to the rate of the reaction according to the rate laws in **Model 2**? Is your answer the same for zero, first, and second order reactions?

10. If the concentration of reactant A increases, what happens to the rate of the reaction, according to the rate laws in **Model 2**? Is your answer the same for zero, first, and second-order reactions?

11. What do you need to plot on the y-axis if you plot time on the x-axis in order to obtain a straight line for each of the reaction orders in **Model 2**? To answer this question, compare the integrated rate laws to the general equation for a straight line, $y = mx + b$, where m is the slope, and b is the intercept on the y-axis when $x = 0$.

    a) a zero-order reaction.

    b) a first-order reaction.

    c) a second-order reaction.

12. What are the slope and y-intercept of the straight line that you identified in your answer to Key Question 11 for:

    a) a zero-order reaction?

    b) a first-order reaction?

    c) a second-order reaction?

EXERCISES

1. On the graph in **Model 1**, draw a vertical line at the half-life time, and label this line $t_{1/2}$. Draw a horizontal line on the graph in **Model 1** at half the initial concentration, and label this line $C_{1/2}$.

2. Compare your answers to Key Question 11 with the graph in **Model 1**, and identify the order of the ozone decomposition reaction.

3. Write the rate law for the ozone decomposition reaction.

4. Write the integrated rate law in both exponential and logarithmic forms for the ozone decomposition reaction, and describe how they are related to the graph in **Model 1**.

5. Identify the order of the decomposition reaction with respect to ozone.

6. Identify the overall order of the ozone decomposition reaction.

7. Determine what would happen to the rate of decomposition if the concentration of ozone were doubled.

8. Determine the rate constant (magnitude and units) for the ozone decomposition reaction.

Got It!

1. Given the following rate law: Rate = k $[CHCl_3]\,[Cl_2]^{1/2}$
   a) Write the reaction order with respect to chloroform.

   b) Write the reaction order with respect to chlorine.

   c) Write the overall reaction order.

   d) Identify what happens to the reaction rate if the concentration of chloroform is cut in half.

   e) Identify what happens to the reaction rate if the concentration of chlorine is increased by four times.

2. Label the following graphs appropriately as zero-order, first-order, or second-order reactions.

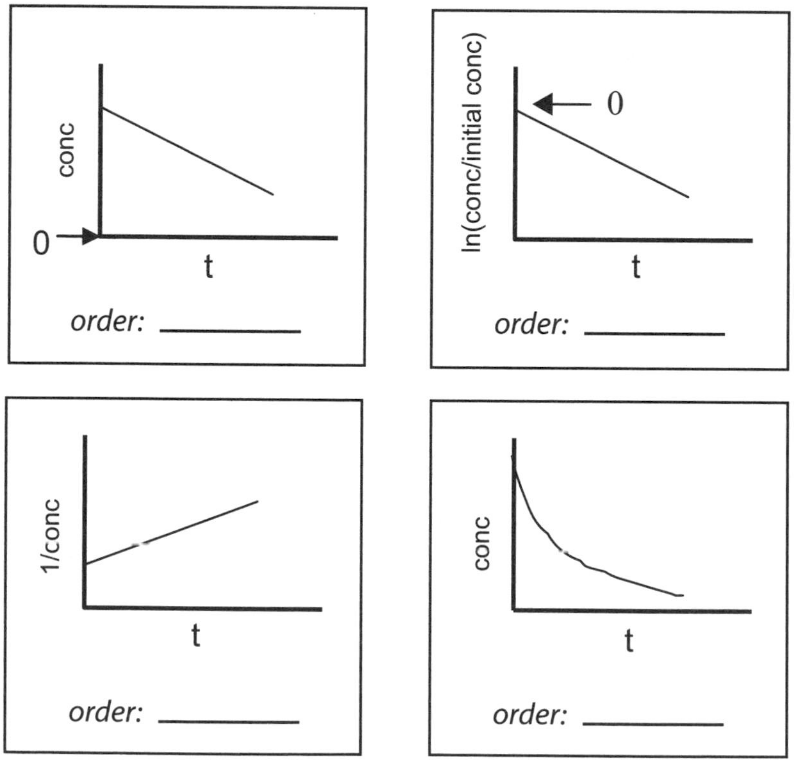

3. The statement, ***the rate decreases as the reaction proceeds***, does not apply to which one of the following?
   a) a zero-order reaction b) a first-order reaction c) a second-order reaction

PROBLEMS

The data in the following table (molar concentration of N_2O_5 versus time) were obtained for the decomposition of dinitrogen pentoxide:

$$2\ N_2O_5(g) \longrightarrow 4\ NO_2(g) + O_2(g)$$

time in s	0	1000	2000	8000	9000	15000	16000	20000	21000
conc in M	0.75	0.65	0.57	0.24	0.21	0.092	0.080	0.046	0.040

1. Determine, for example by making the appropriate graphs, whether this decomposition reaction is zero, first, or second order with respect to N_2O_5.

2. The stoichiometric coefficient for N_2O_5 in the reaction equation is 2. Depending on your answer to Problem 1, explain why the reaction order you obtained in Problem 1 is also 2, or explain why it can be different from 2.

3. Determine the rate constant for this reaction (magnitude and units).

4. Sketch a graph showing the relative concentrations of N_2O_5, NO_2, and O_2 as a function of time as the reaction proceeds. Your graph should show these concentrations on the same scale, and should show the final concentrations relative to each other after equilibrium has been reached.

5. Write a paragraph explaining why the concentrations vary as shown by your graph.

Activity 13-2

Reaction Mechanisms

Why?

A reaction mechanism is a sequence of elementary reactions that describes what we believe takes place as the reactant molecules are transformed into product molecules. A rate law can be predicted from a proposed reaction mechanism, and you can compare predicted and experimental rate laws to determine whether a proposed reaction mechanism is consistent with the experimental rate law. In order for the mechanism to be correct, the predicted rate law must match the rate law that is determined experimentally.

Learning Objectives

- Understand how a chemical reaction proceeds by a series of elementary reactions
- Develop the ability to produce a rate law from experimental data
- Develop the ability to produce a rate law from a proposed reaction mechanism

Success Criteria

- Determine the rate law for a reaction from experimental rate data
- Design a reaction mechanism that produces the experimentally determined rate law and the overall balanced chemical reaction equation

Prerequisite

- **Activity 13-1:** *Rates of Chemical Reactions*

Information

A *chemical reaction mechanism* is a sequence of reactions that depict what is happening at the molecular level. These molecular-level reactions are called *elementary reactions*. The sum of the elementary reactions in a mechanism gives the overall balanced reaction equation.

A rate law can be predicted from a proposed reaction mechanism. If the predicted rate law does not match the experimental rate law, then the proposed mechanism cannot be correct. If the predicted rate law matches the experimental rate law, then it is possible that the proposed mechanism is correct. It is also possible for an incorrect mechanism to produce the correct rate law so a match is not definitive!

An elementary reaction usually involves one or two molecules as reactants. Rarely are more than two molecules involved in an elementary reaction. A *unimolecular* reaction involves only one molecule. A *bimolecular* reaction involves two molecules.

Model 1: Experimental Rate Law for the Oxidation of NO

The rate law for the reaction of nitrogen monoxide with oxygen in the atmosphere has been determined to be $\text{Rate} = -\frac{\Delta[O_2]}{\Delta t} = k[NO]^2[O_2]$ from the following data at 15 °C.

$[NO]_o$ (mol/L)	$[O_2]_o$ (mol/L)	Initial Rate (mol /L min)
0.15	0.15	1.05
0.15	0.30	2.06
0.30	0.30	8.37

Key Questions

1. From the rate law from **Model 1**, what is:
   a) The order of the reaction with respect to oxygen?

   b) The order of the reaction with respect to nitrogen monoxide?

   c) The overall order of the reaction?

2. Based on the data in the table, what happens to the initial rate when the O_2 concentration is doubled while the NO concentration is kept constant? (Use quantitative descriptions in your answer, e.g., decreases by one half, increases by a factor of three, etc.)

3. From the data in the table, what happens to the initial rate when the NO concentration is doubled while the O_2 concentration is kept constant?

4. Are your answers to Key Questions 2 and 3 consistent with the rate law in **Model 1** and the reaction orders with respect to NO and O_2? Describe how you determined whether they are consistent or inconsistent.

5. Based on the insight you gained by answering the Key Questions, how can data such as that given in **Model 1** be used to determine an experimental rate law for a chemical reaction?

Model 2: A Mechanism for the Oxidation of NO

A chemical reaction equation only summarizes the numbers of molecules or moles involved in the overall reaction. For example, the reaction equation for the oxidation of nitrogen monoxide,

$$2NO(g) + O_2(g) \longrightarrow 2NO_2(g),$$

does not mean that two molecules of nitrogen monoxide collide with an oxygen molecule to produce two molecules of nitrogen dioxide. Such a collision between three molecules in the gas phase is extremely improbable, roughly like three cars arriving at an intersection at exactly the same time from three different directions.

The following mechanism has been proposed for this reaction, and it consists of the following elementary reactions:

Step 1 (fast equilibrium): $2\,NO \underset{k_{-1}}{\overset{k_1}{\rightleftharpoons}} N_2O_2$

Step 2 (slow reaction): $N_2O_2 + O_2 \xrightarrow{k_2} 2\,NO_2$

If we add these two steps together, we obtain the overall reaction equation.

$$2\,NO + N_2O_2 + O_2 \longrightarrow N_2O_2 + 2\,NO_2$$

$$2\,NO + O_2 \longrightarrow 2\,NO_2$$

Step 1 involves two elementary reactions. The forward reaction is a bimolecular collision between 2 NO molecules to form N_2O_2. The reverse reaction is the dissociation of N_2O_2, which is a unimolecular reaction. The mechanism proposes that as result of these two reactions, an equilibrium concentration of N_2O_2 is established quickly.

Step 2 is a bimolecular reaction involving the collision of N_2O_2 with O_2. Since the rate of an elementary reaction is proportional to the concentration of the reactants, the rate for the reaction in Step 2 is:

(1) $$\text{Rate} = k_2[N_2O_2][O_2]$$

The mechanism proposes that this rate is slow. The slow step in a reaction is called the *rate-limiting step.*

N_2O_2 is a reaction intermediate. A *reaction intermediate* is a chemical species that appears in an elementary reaction but does not appear in the overall reaction equation. To obtain the experimental rate law, the fast equilibrium in Step 1 is used to eliminate the reaction intermediate N_2O_2

Model 2: A Mechanism for the Oxidation of NO (continued)

rate forward = rate reverse

$$k_1[NO]^2 = k_{-1}[N_2O_2]$$

$$[N_2O_2] = \frac{k_1}{k_{-1}}[NO]^2$$

$$\text{Rate} = \frac{k_2 k_1}{k_{-1}}[NO]^2[O_2]$$

The proposed mechanism therefore produces the experimentally observed rate law. Also adding the elementary reactions produces the balanced reaction equation.

Key Questions

6. How many elementary reactions are proposed in the mechanism for the oxidation of NO in **Model 2**?

7. Which of the elementary reactions in **Model 2** are unimolecular and which are bimolecular?

8. How is the rate law for an elementary reaction determined? For example,

   Rate for Step 2 = $k_2[N_2O_2][O_2]$ and Rate for Reverse Reaction in Step 1 = $k_{-1}[N_2O_2]$

9. In the proposed reaction mechanism, if the forward reaction in Step 1 were the slow step, what would be the rate law for the oxidation of NO?

10. An alternative mechanism for the oxidation of NO is an elementary reaction in which 3 molecules collide to give 2 molecules of NO_2.

    $$2NO + O_2 \longrightarrow 2NO_2$$

    a) What is the rate law predicted by this reaction mechanism?

    b) Is it consistent with the experimental rate law identified in **Model 1**?

c) Can the experimental rate law serve to identify whether this mechanism or the mechanism in **Model 1** is not correct?

d) Why would you not expect a collision involving three molecules to be the mechanism for a reaction?

11. As a general rule, can the rate law uniquely identify the reaction mechanism or can it only serve to eliminate incorrect mechanisms? Explain.

Exercises

1. Given the data in the table for **Model 1**, determine the value and units for the rate constant for the oxidation of nitrogen monoxide.

2. Show that the sum of the two steps for the mechanism in **Model 2** yields the overall reaction equation.

3. Show that the experimental rate law for the oxidation of NO in **Model 1** can be obtained from the mechanism proposed in **Model 2** and that the intermediate N_2O_2 does not appear in the rate law expression.

PROBLEMS

1. Develop a two-step mechanism involving oxygen atoms in the oxidation of nitric oxide and determine the rate law that it predicts. Note that oxygen atoms are very reactive and will react quickly.

2. Use the kinetic data in the table below to determine the rate law for the following reaction:

$$aA + bB + cC \longrightarrow dD + eE$$

Initial Concentrations (mol/L)			Initial Rate (mol/L s)
[A]	[B]	[C]	
0.20	0.20	0.10	2.67
0.10	0.20	0.20	1.33
0.30	0.05	0.10	0.25
0.10	0.05	0.10	0.08
0.10	0.10	0.20	0.33
0.10	0.10	0.10	0.33

3. The following mechanism is proposed for the reaction of chlorine with chloroform:

$$Cl_2 + CHCl_3 \longrightarrow CCl_4 + HCl$$

Step 1 fast equilibrium $\quad Cl_2 \underset{k_{-1}}{\overset{k_1}{\rightleftarrows}} 2\,Cl$

Step 2 slow $\quad CHCl_3 + Cl \xrightarrow{k_2} HCl + CCl_3$

Step 3 fast $\quad CCl_3 + Cl \xrightarrow{k_3} CCl_4$

a) Show that this mechanism gives the correct, overall balanced reaction equation.

b) Write the rate law predicted by this mechanism.

Activity 13-3

Activation Energy and Catalysis

Why?

Not all collisions between molecules result in a reaction. The molecules must have enough energy for the reaction to occur. The minimum energy that is needed is called the *activation energy*. Catalysts are substances that lower the activation energy and thereby increase the rate of reactions. Catalysts are essential in the production of industrial chemicals. Biological catalysts, which are called *enzymes*, are essential for life and for the development of new pharmaceutical products.

Learning Objectives

- Understand the factors that limit the rate of a chemical reaction
- Be able to determine the activation energy of a chemical reaction from reaction rate data
- Recognize how catalysts can increase reaction rates

Success Criteria

- Ability to produce a complete list of factors that affect the rate of a chemical reaction
- Ability to identify three or more ways that the rate of a chemical reaction can be increased
- Correct determination of activation energies from reaction rate data
- Correct quantitative prediction of how large a rate increase will be produced by a given change in activation energy

Prerequisites

- **Activity 13-1:** *Rates of Chemical Reactions*
- **Activity 13-2:** *Reaction Mechanisms*

Information

For a reaction to occur, molecules must collide. The frequency of the collisions affects the rate of the reaction. The frequency can be changed by increasing or decreasing the concentrations of the reactants, and by changing the temperature to change the velocities of the molecules.

But even when molecules collide, they do not all react. In order to react, the two molecules must be oriented in just the right way. Unfortunately, there is nothing that can be done to control molecular orientations, except in very sophisticated experiments.

Molecules must also have enough energy for the reaction to occur. Some minimum energy is needed because existing bonds must be broken and new bonds must be formed. The point in the reaction at which this reconstruction is occurring is called the *transition state* or *activated complex*, and the energy needed to reach the transition state is called the *activation energy*, E_a. Only those molecules that have this amount of energy or more will react and produce products. The energy of the molecules can be changed by raising or lowering the temperature.

Model 1: Energy vs Reaction Coordinate Diagram

An energy vs reaction coordinate graph is used to show how the potential energy of the reactants changes as the reactants turn into products. The reaction coordinate is a measure of the progress of the reaction along the reaction pathway.

Figure 1

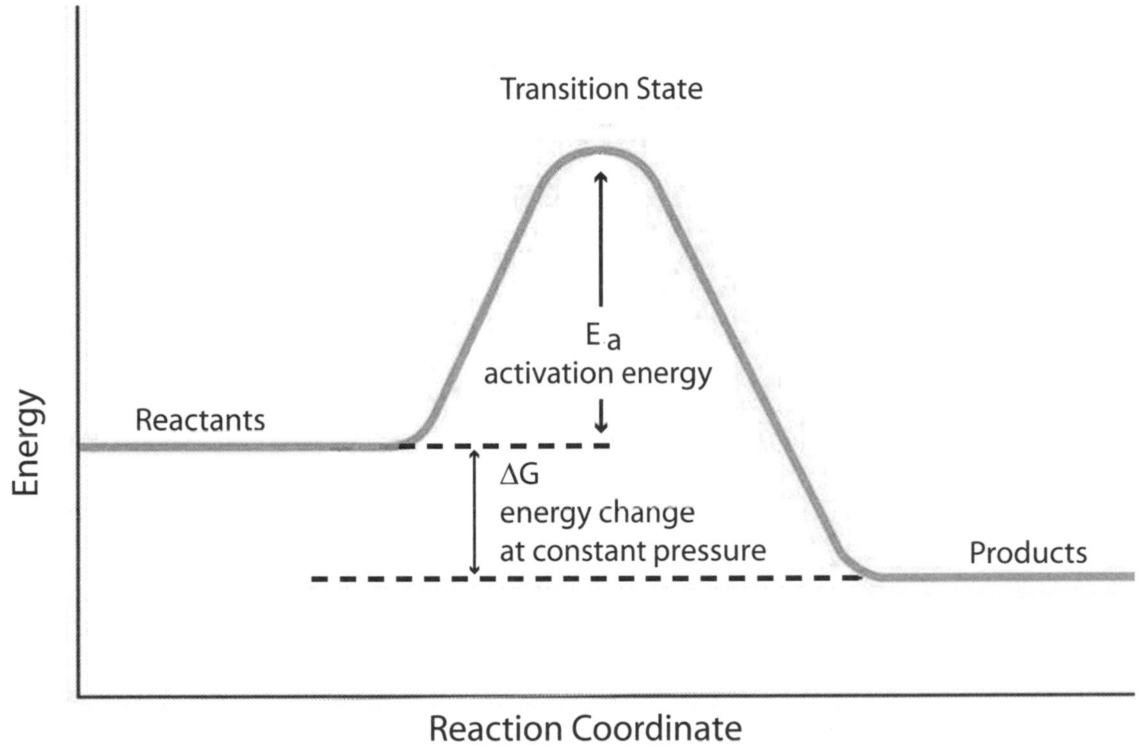

Key Questions

1. Is the energy change in going from reactants to products in **Model 1** positive or negative?

2. According to **Model 1**, which has the higher energy, the reactants or the transition state?

3. Which molecules are more likely to reach the transition state and pass over to products when they collide, those with high velocities and kinetic energies or those with low velocities and kinetic energies?

4. As the temperature increases, does the fraction of molecules with high kinetic energies increase or decrease?

5. In view of **Model 1** and your answers to Key Questions 1-4, why do you think the rate of a chemical reaction increases with increasing temperature?

6. Draw an arrow on the diagram in **Model 1** to indicate the magnitude of the activation energy for the reverse reaction (products going back to reactants). Do you think the rate constant of the reverse reaction will be larger or smaller than that of the forward reaction? Explain your answer in terms of the fraction of molecules that have enough kinetic energy to reach the transition state.

EXERCISES

1. Hydrogen and chlorine react to produce hydrochloric acid, but the reverse reaction also occurs at a slower rate. Consider the reaction $2HCl \longrightarrow H_2 + Cl_2$. Draw two diagrams, one showing an orientation of the two HCl molecules that is unfavorable for this reaction, and one showing an orientation that is favorable for this reaction.

2. Give two or more reasons why some collisions between molecules might not result in a chemical reaction.

3. Draw an energy vs reaction coordinate diagram to illustrate a reaction in which the energy of the products is greater than the energy of the reactants. Label all quantities as in **Model 1**.

4. Using your diagram from Exercise 3, identify which reaction (forward or reverse) has the larger activation energy and which has the larger rate constant.

INFORMATION

Because a reaction is faster at higher temperatures, the rate constant for the reaction must be larger. The activation energy for a reaction can therefore be determined from experimental measurements of the rate constant at several temperatures. When such data are plotted in the form ln(k) vs 1/T as shown in Figure 2, a straight line is produced. This result means that the rate constant varies exponentially with 1/T as given by the Arrhenius equation.

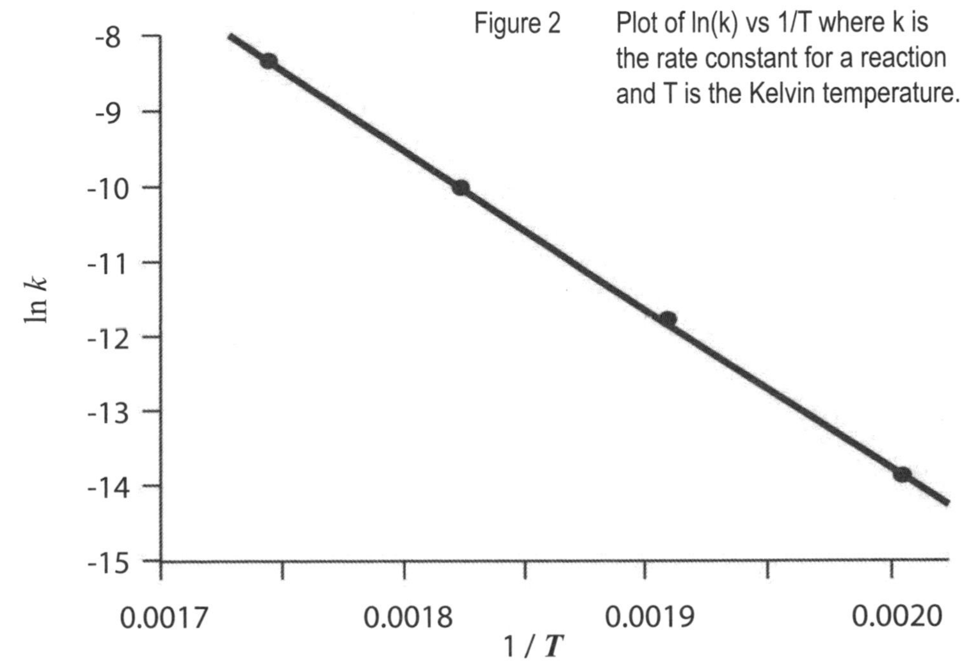

Figure 2 Plot of ln(k) vs 1/T where k is the rate constant for a reaction and T is the Kelvin temperature.

Model 2: The Arrhenius Equation

$$k = Ae^{-\frac{E_a}{RT}}$$

A is the frequency factor. The frequency factor depends on how often molecules collide when the concentrations are 1M, and on whether the molecules are properly oriented when they collide. A is the value that the rate constant would have if all the molecules had enough energy to react, e.g., when the activation energy $E_a = 0$ or the Kelvin temperature T is very large.

Key Questions

7. Since the gas constant R has units of J/mol K, what are the units of E_a in the Arrhenius equation?

8. Does the Arrhenius equation predict that the rate constant will increase or decrease if the activation energy gets larger?

9. Does the Arrhenius equation predict that the rate constant will increase or decrease as the temperature increases?

Exercises

5. Show that the Arrhenius equation produces the following equation by taking the natural logarithm of both sides and using the property ln(ab) = ln(a) + ln(b).

$$\ln k = \ln A - \frac{E_a}{R}\frac{1}{T}$$

6. The equation in Exercise 5 corresponds to that of a straight line when ln(k) is plotted versus 1/T. Identify the quantities in the equation that determine the slope of the line and the quantity that determines the intercept at 1/T = 0.

7. The line in Figure 2 is described by $\ln(k) = 28.5 - (2.1 \times 10^4\ \text{K})/T$.

   a) Determine the frequency factor and the activation energy for the reaction.

   b) The rate constant for a second order reaction has units of L $\text{mol}^{-1}\ \text{s}^{-1}$. What are the units of the frequency factor, A, for a second order reaction?

PROBLEMS

1. Fireflies flash at a rate that is temperature dependent. At 29 °C the average firefly flashes at a rate of 3.3 flashes every 10 seconds. At 23 °C the average rate is 2.7 flashes every 10 seconds. Use the Arrhenius equation to determine the activation energy for the flashing process.

INFORMATION

A catalyst, as shown in **Model 3**, changes the mechanism of a chemical reaction and lowers its activation energy. The catalyst participates in intermediate steps of the reaction, but it is neither produced nor consumed in the reaction so the balanced reaction equation remains the same.

Model 3: Catalysts

Figure 3

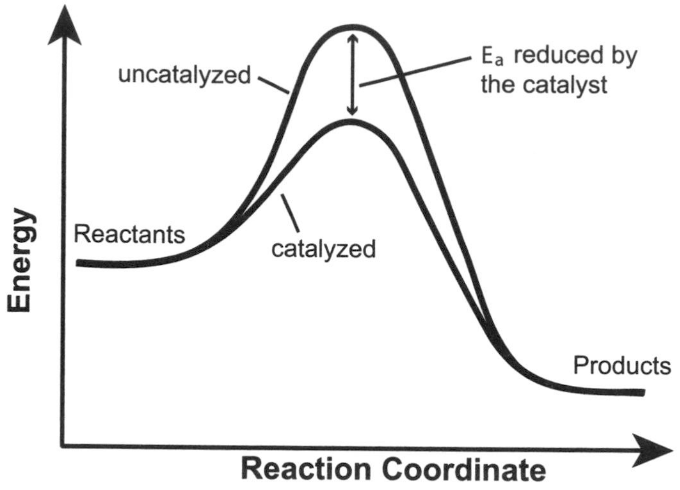

Key Questions

10. What effect does a catalyst have on the activation energy of a reaction?

11. What effect does a catalyst have on the change in free energy of a reaction?

12. What effect does a catalyst have on the mechanism of a reaction?

13. What effect does a catalyst have on the stoichiometry of a reaction?

14. How does the rate of the rate limiting step in a reaction with a catalyst compare with the rate of the rate limiting step without the catalyst?

15. What are at least three ways that the rate of a chemical reaction can be increased?

PROBLEMS

2. A catalyst reduces the activation energy for a reaction from 17 kJ/mol to 2 kJ/mol. By what factor is the rate accelerated? Assume that the frequency factor A does not change, and that the temperature is 200 K.

Activity 14-1

Spontaneous Change and Entropy

Why?

Sometimes things seem to happen without outside intervention. Such changes are called *spontaneous*. Spontaneous change occurs because things move to a more probable situation. The concept of entropy quantifies the probability of a situation and enables you to predict whether or not a change or a chemical reaction will be spontaneous.

Learning Objective

- Understand the meaning of spontaneous change and entropy

Success Criterion

- Ability to correctly identify whether entropy increases or decreases

Vocabulary

- *Spontaneous change* happens due to a natural tendency without an apparent external cause.
- *Entropy* is a measure of the dispersal of energy over the states available to a system.

Information

There is a natural tendency for energy to disperse. Entropy provides a quantitative measure of the extent to which energy has dispersed. To *disperse* means "to spread out." Dispersal of energy occurs because there are more ways for the energy to be spread out over many states than there are for it to be concentrated in one or a few states. You can see this dispersal occurring when a hot object that is surrounded by air cools. The energy is in the hot object initially. As it cools, the energy is dispersed into the surrounding air molecules. The object cools down spontaneously. It does not heat up spontaneously. The reason for this is that there are many more ways for the energy to be distributed into the surrounding air molecules than there are for it to be distributed within the object.

Entropy is related to the number of states that are accessible with a given amount of energy. The larger the number of states, the more dispersed the energy can be. A system that can be found in a large number of states often appears to be more disordered than a system that is always found in only one state. Consequently entropy often is associated with the idea of order and disorder, but its real significance lies in its association with energy dispersal and the number of accessible states.

Model: Discover Entropy on Your Bookshelf!

Entropy has a connection with the concepts of probability and the number of accessible states. To see this connection, consider the number of arrangements for three books on a shelf. A particular arrangement of books on a shelf is defined to be a state. An arrangement with a large number of ways of occurring will be more probable than an arrangement with only a single possibility.

Arrangement Set	Constraint	Number of States
A	Three books are lined up alphabetically by author starting on the left with the binding facing outward and the titles on the binding facing left.	1
B	The books do not need to be arranged alphabetically by author, but the bindings face outward and the titles face left.	6
C	The books do not need to be arranged alphabetically by author; the bindings must face outward, but the titles can face right and left.	48
D	The books do not need to be arranged alphabetically by author; the bindings can face outward, inward, up, or down; and the titles can face right and left.	3072

Key Questions

1. What is the maximum number of states of the three books in the Model?

2. Which arrangement set has the highest probability: A, B, C, or D?

3. Which arrangement set would you consider to be the most disordered?

4. Which arrangement set would you say has the highest entropy?

5. If you begin the semester with arrangement A, in which arrangement set will the state of your bookshelf be at the end of the semester, if you use your books and allow your the order of your bookshelf to change spontaneously during the course of the semester (i.e., without extra care, effort, and intervention on your part)?

6. Would you expect to get the most ordered arrangement by spontaneous change? Explain in terms of the number of states and the probability of the most ordered arrangement.

7. Based on the Model, what connections do you see between the number of accessible states, probability, entropy, and disorder?

8. Based on the Model, what connection do you see between entropy and spontaneous change?

9. Identify another example of a spontaneous change accompanied by an increase in entropy.

INFORMATION

An increase or decrease in entropy accompanying various physical and chemical changes can be identified from the following guidelines. Entropy increases when particles can move around more freely and access a larger number of states associated with this motion. When a gas expands into a larger volume, the number of potential states (as specified by the position and momentum of each particle) available to the particles increases, as does the entropy. Consequently, the entropy increases when a chemical reaction produces a gas phase product from a solid or when a gas phase molecule dissociates into two or more particles. Similarly, when a solid melts or a liquid vaporizes, the particles have more positions available to them and the entropy increases.

The decrease in molecular motion that accompanies precipitation would be expected to reduce the number of accessible states and decrease the entropy. Exceptions occur, however, because solvent molecules can bind tightly to ions when they form solutions. Consequently the decreased entropy of the precipitating solute is counteracted by a larger increase in entropy of the solvent. Thus, for example, the entropy change is -33 J mol^{-1} K^{-1} for precipitation of $AgCl$, but $+115$ J mol^{-1} K^{-1} for precipitation of $MgCl_2$. There is no general rule for the entropy change on dissolving and precipitating when both the solute and solvent are taken into consideration.

Exercises

Identify whether the entropy increases or decreases for each of the following changes and explain why.

a) evaporation of a liquid

b) freezing of a liquid

c) precipitation of a solid from solution

d) increase in the volume of a gas at constant temperature

e) $CaCO_3(s) \longrightarrow CaO(s) + CO_2(g)$

f) $N_2(g) + 3H_2(g) \longrightarrow 2NH_3(g)$

Research

Explain how the number of states for each book arrangement set in the Model was determined.

ACTIVITY 14-2

Entropy of the Universe and Gibbs Free Energy

WHY?

While entropy can be used to predict whether or not a process is spontaneous, one must consider the total entropy, i.e., the entropy of the universe, to do so accurately. This consideration leads to the concept of *free energy*, which is easier to apply because it deals only with the system being studied, not the universe. The concept of free energy is very useful. For example, you can use it to determine the concentrations of reactants and products in a chemical reaction at equilibrium, the voltage produced by batteries, and whether or not a chemical reaction will occur spontaneously.

LEARNING OBJECTIVE

- Understand the relationship between entropy and free energy

SUCCESS CRITERIA

- Correct prediction of increases and decreases in entropy and free energy
- Correct identification of spontaneous processes
- Accurate calculations of entropy and free energy changes

PREREQUISITES

- **Activity 06-2:** *Internal Energy and Enthalpy*
- **Activity 14-1:** *Spontaneous Change and Entropy*

INFORMATION

Free energy combines the ideas of enthalpy and entropy. Physically, the free energy change is the maximum work that can be done by a spontaneous process at constant temperature and pressure. If the process is not spontaneous, then the free energy is the minimum work needed to make the process occur.

Entropy, S, can be defined in terms of the number of states, W, that are accessible to a system with a given amount of energy:

$$S = k \ln(W) \quad \text{where } k = \text{Boltzmann's constant}$$

The change in entropy, ΔS, can be defined in terms of the change in the number of accessible states or in terms of the heat flow, q, into or out of a system at constant temperature:

$$\Delta S = q/T$$

The diagram in **Model 1** depicts heat flowing from a system into the surroundings. The system plus the surroundings represents the universe. The change in energy of the system is given by $\Delta E = q + w$, where q is the heat added to the system and w is the work done on the system. Since the heat is leaving the system in the model, it is given a negative sign, $-q_p$. The change in enthalpy of the system is also negative.

$$\Delta H = H_{final} - H_{initial} = -q_p$$

The subscript P means that the pressure is constant.

Model: Change in the Entropy of the Universe

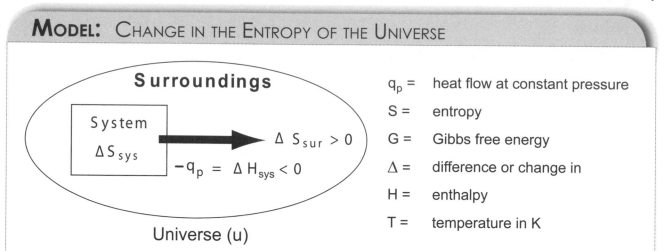

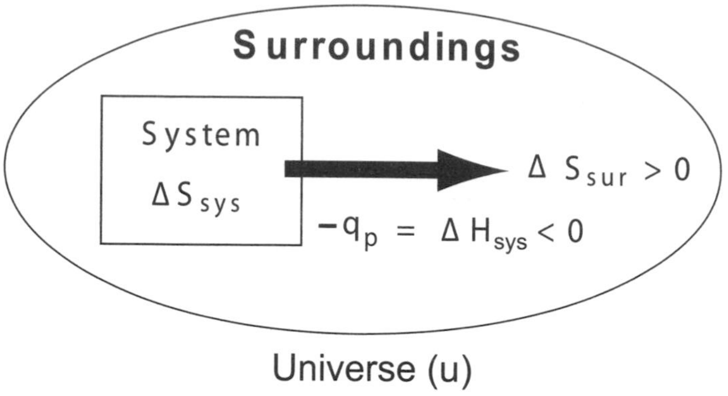

q_p =	heat flow at constant pressure
S =	entropy
G =	Gibbs free energy
Δ =	difference or change in
H =	enthalpy
T =	temperature in K

The entropy change of the universe is the total entropy change of the system and its surroundings.

(1) $\Delta S_u = \Delta S_{sur} + \Delta S_{sys}$

Heat flowing into the surroundings increases the entropy of the surroundings. Since ΔH_{sys} is negative for an exothermic process, the following equation requires a minus sign to make ΔS_{sur} positive, corresponding to an increase.

(2) $\Delta S_{sur} = -\Delta H_{sys} / T$

Substituting equation (2) into equation (1) and multiplying through by –T gives equation (3).

(3) $-T\,\Delta S_u = \Delta H_{sys} - T\,\Delta S_{sys}$

The quantity on the right side, which depends only on the system, is defined as the free energy of the system, or more specifically the Gibbs free energy.

(4) $\Delta G_{sys} = \Delta H_{sys} - T\,\Delta S_{sys}$

The free energy is also related to the entropy change of the universe by equation (5)

(5) $\Delta G_{sys} = -T\,\Delta S_u$

The free energy accounts for the entropy of the universe in terms of the enthalpy and entropy of the system. These quantities (ΔG_{sys}, ΔH_{sys}, ΔS_{sys}) all refer to the system so the subscript *sys* is not necessary and will no longer be used.

Key Questions

1. A process occurring in a system can cause the entropy of the surroundings to change. In the Model, what is transferred to the surroundings to cause the entropy of the surroundings to change?

2. When $\Delta H < 0$:

   a) does the energy of the system increase or decrease?

   b) does heat flow into or out of the surroundings?

   c) does the entropy of the surroundings increase or decrease?

3. According to the Model, what do you need to know in order to calculate the entropy change in the surroundings?

4. What is the definition of the free energy change of the system in terms of the entropy of the universe?

5. For a process to be spontaneous, what must happen to the entropy of the universe?

6. For a process to be spontaneous, what must happen to the free energy of the system in which the process is occurring?

7. Why can a change be spontaneous even if the system entropy decreases?

8. What is an example of a spontaneous change in which the entropy of the system decreases?

Exercises

1. Show how the following key equation was derived in the Model: $\Delta G = \Delta H - T\,\Delta S$

2. Calculate the change in entropy of the surroundings when 1.0 mol of water vaporizes at 100 °C. The heat of vaporization of water is 40.6 kJ/mol.

3. Calculate the free energy change for the following reaction and predict whether the reaction occurs spontaneously at 25 °C.

$$H_2(g) + CO_2(g) \longrightarrow H_2O(g) + CO(g)$$

$$\Delta H = 41.8 \text{ kJ}, \quad \Delta S = 42.1 \text{ J/K}$$

4. The table below shows all the possible combinations of signs for the entropy and enthalpy changes in any process. Enter a +, –, or T in the ΔG column to correspond to these changes. Use T for temperature if the sign of ΔG depends on the temperature. If there is a temperature effect, specify whether the reaction is spontaneous at a high or a low temperature.

ΔS	ΔH	ΔG + or – or T	Spontaneous? enter yes or no or specify if it is spontaneous at a high T or low T
+	–		
–	+		
+	+		
–	–		

5. Is an exothermic reaction or process always spontaneous? Explain.

6. Can an endothermic reaction or process be spontaneous? Explain.

7. Generally when biological enzymes are heated, they lose their catalytic activity. The process is endothermic and spontaneous. Is the structure of the active enzyme more or less ordered than the inactive enzyme? Explain.

Activity 15-1

Dynamic Equilibrium and Le Châtelier's Principle

Why?

When opposing forces or issues are balanced, a system is said to be at equilibrium. In chemical reactions, equilibrium is dynamic because the forward and reverse reactions are occurring continuously. Equilibrium is reached when the rate of the reverse reaction equals the rate of the forward reaction. Le Châtelier's Principle provides a simple qualitative rule for analyzing the effects of perturbations on equilibrium situations. You can use this information in evaluating the extent of a chemical reaction, predicting the effects of changes, and determining concentrations of reactants and products.

Learning Objectives

- Understand the meaning and implications of dynamic equilibrium
- Discover how systems will change to reach equilibrium

Success Criteria

- Correctly apply the idea of dynamic equilibrium in particular situations
- Quality of predictions regarding how a system will change to reach equilibrium

Prerequisites

- **Activity 04-2:** *Dissociation and Precipitation Reactions*
- **Activity 04-3:** *Introduction to Acid – Base Reactions*
- **Activity 06-2:** *Internal Energy and Enthalpy*
- **Activity 11-3:** *Partial Pressure*
- **Activity 13-3:** *Activation Energy and Catalysis*

Model 1: Dynamic Equilibrium

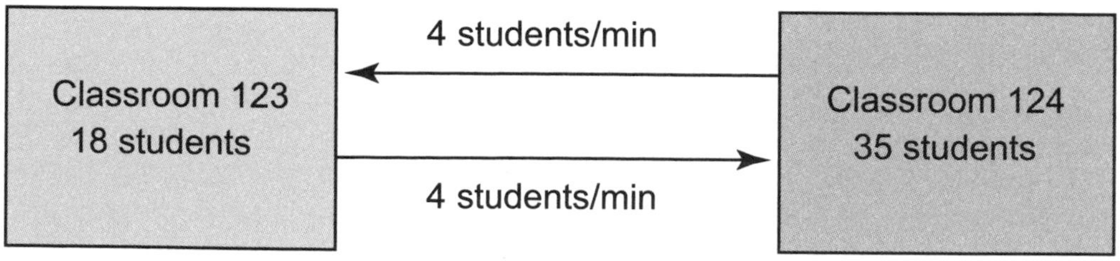

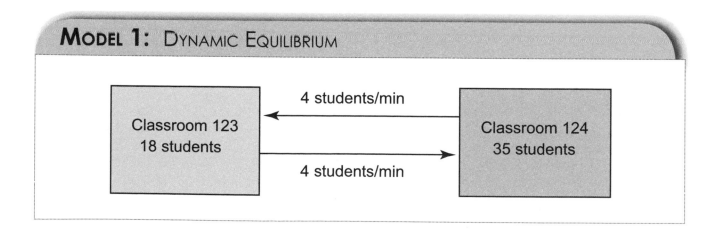

KEY QUESTIONS

1. Are the students moving from room to room in the model?

2. Is the number of students in each room changing?

3. In the same way that equilibrium is reached for the students in the classrooms, chemical equilibrium is reached in reactions when (select one):

   a) all reactions stop

   b) the forward reaction stops

   c) the concentrations of products and reactants become equal

   d) the rates of the forward and reverse reactions become equal

   e) the temperature becomes constant

4. Are the concentrations of A, B, C, and D changing if 100 molecules per second of A and B are being converted into C and D by the forward reaction and 100 molecules per second of C and D are being converted into A and B by the reverse reaction? Explain.

5. What is an example of dynamic equilibrium that you have seen?

EXERCISES

1. After the following reaction has reached equilibrium, all the carbon in CO is magically converted to the isotope carbon-14. Will the carbon-14 always be found only in the CO or will some be found in the CO_2? Explain.

$$H_2O(g) + CO(g) \rightleftarrows H_2(g) + CO_2(g)$$

2. Consider this reaction: $H_2O + CO \rightleftarrows H_2 + CO_2$

   In one experiment 1.0 mole H_2O and 1.0 mole CO are put into a flask and heated to 350 °C. In a second experiment 1.0 mole H_2 and 1.0 mole CO_2 are put into another flask with the same volume as the first and heated to 350 °C. After equilibrium is reached in both cases, is there any difference in the composition of the mixtures in the two flasks? Explain.

Model 2: Le Châtelier's Principle

> If a change is imposed on a system, the position of equilibrium will shift in a direction that tends to reduce the effect of that change.

Key Questions

6. What will happen to the number of students in the two rooms in **Model 1** if the temperature in Classroom 124 rises because the air exchange system for that room fails? Explain your answer in terms of Le Châtelier's Principle.

7. According to Le Châtelier's Principle, what will happen in a chemical reaction at equilibrium if a reaction product is removed?

8. According to Le Châtelier's Principle, what will happen in a chemical reaction at equilibrium if more reactant is added?

9. What is an example of Le Châtelier's Principle applied in an area other than chemistry?

EXERCISES

3. An important endothermic gas phase reaction for the commercial production of hydrogen is

$$CH_4(g) + H_2O(g) \rightleftarrows 3\ H_2(g) + CO(g)$$

Use Le Châtelier's Principle to explain what will happen to the number of moles of hydrogen at equilibrium when

a) Carbon monoxide is removed.

b) Water vapor is removed.

c) Methane is added.

d) Carbon monoxide is added.

e) The pressure is increased by adding argon gas, but the partial pressures of the reactants and products do not change.

f) The pressure is decreased by opening a valve to a second reaction vessel, which lowers the partial pressures of the reactants and products.

g) A catalyst is added. Catalysts change the mechanism for a reaction.

h) The temperature is increased.

Problems

1. Nitric oxide, a significant air pollutant, is formed from nitrogen and oxygen at high temperatures, e.g., in an automobile engine, by the following reaction:

$$N_2(g) + O_2(g) \rightleftarrows 2\ NO(g)$$

As the piston compresses these gases in the cylinder of an automobile engine, what happens to the equilibrium amount of NO in the cylinder? In compressing the gases, the piston moves in the cylinder to reduce the volume and increase the pressure. Identify whether the amount of NO in the cylinder increases, decreases, or stays the same. Explain why.

2. The synthesis of ammonia from nitrogen and hydrogen is an important industrial process for producing nitrates used in fertilizer and explosives. The reaction is exothermic:

$$N_2(g) + 3H_2(g) \rightleftarrows 2NH_3(g)$$

To obtain the highest yield of ammonia at equilibrium, what combination of temperature and pressure would you use? Explain.

a) high T and P b) low T and P c) high T, low P d) low T, high P

Activity 15-2

The Reaction Quotient & Equilibrium Constant

Why?

All chemical reactions, and many other processes as well, eventually reach equilibrium, at which point changes are no longer manifest. For chemical reactions this situation is described by a characteristic constant called the *equilibrium constant*. You can use the equilibrium constant to predict the amount of reactants that will become products and to calculate the equilibrium concentrations of reactants and products under various conditions.

Learning Objectives

- Learn how to write the reaction quotient and equilibrium constant expressions
- Determine how the reaction quotient can be used to predict the direction of a chemical reaction

Success Criteria

- Accurately calculate values for reaction quotients and equilibrium constants
- Correct predictions of reaction direction to reach equilibrium

Prerequisites

- **Activity 04-1:** *Balanced Chemical Reaction Equations*
- **Activity 11-1:** *The Ideal Gas Law*
- **Activity 11-3:** *Partial Pressure*
- **Activity 15-1:** *Dynamic Equilibrium and Le Châtelier's Principle*

Model: The Reaction Quotient and Equilibrium Constant

For a general balanced reaction equation,

$$a\,A + b\,B \rightleftarrows c\,C + d\,D$$

the Reaction Quotient Q is written as

$$Q = \frac{[C]^c[D]^d}{[A]^a[B]^b}$$

where the quantities in square brackets are the molar concentrations divided by the standard concentration of 1M, so the units cancel. The concentrations of the products are in the numerator and the concentrations of the reactants are in the denominator, each raised to the power given by the stoichiometric coefficient.

When the concentrations at equilibrium are used, the Reaction Quotient becomes the Equilibrium Constant K. The Equilibrium Constant characterizes a reaction that is at equilibrium at a given temperature and is given by:

$$K = \frac{[C]_{eq}^c[D]_{eq}^d}{[A]_{eq}^a[B]_{eq}^b}$$

Model: The Reaction Quotient and Equilibrium Constant (con't)

Example:

$$2NOCl(g) \rightleftarrows 2NO(g) + Cl_2(g)$$

At equilibrium the concentrations are

Cl_2 0.030 M NO 0.023 M NOCl 1.0 M

$$K = \frac{[NO]_{eq}^2[Cl_2]_{eq}}{[NOCl]_{eq}^2}$$

$$K = \frac{(0.023)^2(0.030)}{(1.0)^2}$$

$$K = 1.6 \times 10^{-5}$$

Key Questions

1. What items go into the numerator in the reaction quotient and equilibrium constant expressions?

2. What items go into the denominator in the reaction quotient and equilibrium constant expressions?

3. What determines the exponent for each item in the reaction quotient and equilibrium constant expressions?

4. Under the following conditions, which will be present at equilibrium in the largest amounts: reactants or products?
   a) The equilibrium constant is small.
   b) The equilibrium constant is large.

5. If Q is larger than K, will the reaction proceed in the forward direction or the reverse direction to reach equilibrium? Explain your reasoning.

6. If Q is smaller than K, will the reaction proceed in the forward direction or the reverse direction to reach equilibrium? Explain your reasoning.

7. When will Q = K?

Exercises

1. Write the equilibrium constant expression for the reaction

$$2\ N_2O_5\,(g) \rightleftarrows 4\ NO_2\,(g) + O_2\,(g)$$

2. Write the equilibrium constant expression for the following reaction. Note that the concentration of a pure solid is a constant. The moles per liter do not change during the reaction; the solid is just consumed. Consequently the concentration of the solid is not included in the equilibrium constant expression.

$$C(s) + CO_2\,(g) \rightleftarrows 2\ CO(g)$$

3. For reactions in solution, molar concentrations are usually used in equilibrium constant expressions (designated by K or K_c). In gases, partial pressures can also be used (designated by K_p). Equilibrium partial pressures of NOCl, NO, and Cl_2 in a container at 300 K are 1.2 atm, 0.050 atm, and 0.30 atm, respectively. Calculate a value for the equilibrium constant K_p for the following reaction.

$$2\ NO(g) + Cl_2\,(g) \rightleftarrows 2\ NOCl(g)$$

Got It!

Hydrogen iodide decomposes to give hydrogen and iodine. The equilibrium constant is about 2.0×10^{-3} at some temperature. If you were to place some hydrogen iodide in a container and seal the container, what would you expect to find at equilibrium: mostly hydrogen iodide or mostly hydrogen and iodine? Explain how the value of the equilibrium constant helped you answer this question.

Activity 15-3

Free Energy and Chemical Equilibrium

Why?

The free energy change for a chemical reaction depends on the temperature and the concentrations of the reactants and the products. Like chemists in research and manufacturing, you can use the free energy change to identify suitable temperatures for reactions, to determine the direction a reaction will go to reach equilibrium, and to obtain values for equilibrium constants and equilibrium concentrations.

Learning Objectives

- Understand the relationship between the free energy change and the reaction quotient expression
- Identify the effect of temperature on the free energy change and on equilibrium constants

Success Criteria

- Calculate the free energy change for a reaction at various temperatures and various concentrations of reactants and products
- Identify the direction in which a reaction will spontaneously proceed
- Calculate one or more of $\Delta G°$, $\Delta H°$, $\Delta S°$, and, given the others, calculate K

Prerequisites

- **Activity 06-3:** *Hess's Law: Enthalpy is a State Function*
- **Activity 14-2:** *Entropy of the Universe and Gibbs Free Energy*
- **Activity 15-2:** *The Reaction Quotient and Equilibrium Constant*

Information

The change in free energy of a chemical reaction under standard conditions is called the *standard free energy change*, $\Delta G°$. Conditions are "standard" when the pressure is 1 atm, and all reactants and products are either pure solids, liquids, or gases, or are solutes at 1 M concentration.

The free energy change in a chemical reaction (ΔG_T), at some Kelvin temperature T, is related by the following equation to the standard free energy change at that temperature, $\Delta G_T°$, and the reaction quotient, Q, which involves the concentrations of reactants and products, where R is the ideal gas constant:

$$\Delta G_T = \Delta G_T^{\circ} + RT\ln(Q)$$

When reactants and products are in their standard states, then Q is 1 because the concentrations are 1 M. Then $\ln(Q) = \ln(1) = 0$, and $\Delta G_T = \Delta G_T°$ and the change in the free energy is just the standard free energy change.

An increase in the entropy of the universe or a decrease in the free energy of the system is the driving force behind spontaneous change. Concentrations in a chemical reaction change spontaneously to lower the free energy. Equilibrium is reached in a chemical reaction when the free energy can no longer decrease because the free energy of the reactants is equal to the free energy of the products. The free energy change for the reaction, ΔG_T, is then 0.

Since Q = K at equilibrium,	$\Delta G_T = \Delta G_T^\circ + RT\ln(K) = 0$
which means that:	$\Delta G_T^\circ = -RT\ln(K)$
which can be rearranged to provide an expression for K:	$K = e^{-\frac{\Delta G^\circ}{RT}}$
and also, since $\Delta G^\circ = \Delta H^\circ - T\Delta S^\circ$,	$\Delta H^\circ - T\Delta S^\circ = -RT\ln(K)$

Key Questions

1. According to the information provided, what is the relationship between the free energy change for a chemical reaction and the concentrations of the reactants and products?

2. What makes the free energy change (ΔG) for a chemical reaction different from the standard free energy change (ΔG°) for that reaction?

3. According to the information provided, what is the free energy change for a chemical reaction or any other process at equilibrium?

4. Will the chemical reaction with the smaller positive standard free energy change have the larger or smaller equilibrium constant? Explain why your answer is reasonable or makes sense.

5. According to the information provided, why are equilibrium constants for a chemical reaction temperature dependent?

EXERCISES

1. Starting with $\Delta G_T = \Delta G_T^o + RT\ln(Q)$ show that $\Delta G_T^o = -RT\ln(K)$ when reactants and products in a chemical reaction are present at their equilibrium concentrations.

2. Use the equations in the information section to show that the free energy change for a chemical reaction, when all the reactant and product concentrations are 1 M, is just the standard free energy change for that reaction.

3. Methanol is a proposed alternative fuel. The change in standard free energy for the synthesis of liquid methanol from gaseous carbon monoxide and hydrogen is –29 kJ/mole of methanol. Write the balanced reaction equation, the equilibrium constant expression, and determine the value of the equilibrium constant at 25 °C.

4. Calculate the equilibrium constant for the synthesis of ammonia at 350 °C, given that $\Delta H° = -91.8$ kJ mol^{-1} and $\Delta S° = -198.1$ J mol^{-1} K^{-1}

$$N_2(g) + 3H_2(g) \rightleftarrows 2NH_3(g)$$

5. Given that $\Delta G° = 2.8$ kJ mol^{-1}, calculate the free energy change for the decomposition of dinitrogen tetroxide at 298 K when $P_{N2O4} = 5$ atm and $P_{NO_2} = 1$ atm

$$N_2O_4(g) \rightleftarrows 2NO_2(g)$$

6. Calculate the equilibrium constant at 298 K for the reaction of carbon monoxide and hydrogen, $CO(g) + 2H_2(g) \rightleftarrows CH_3OH(l)$, given the following information:

$$\Delta G°_f\, CO(g) = -137.2 \text{ kJ mol}^{-1}$$

$$\Delta G°_f\, H_2(g) = 0 \text{ kJ mol}^{-1}$$

$$\Delta G°_f\, CH_3OH(l) = -166.6 \text{ kJ mol}^{-1}$$

7. Which graph below best describes the variation of ΔG° with temperature when plotting $y = \Delta G°$ and $x = T$, and when both ΔH° and ΔS° are positive? Explain your reasoning.

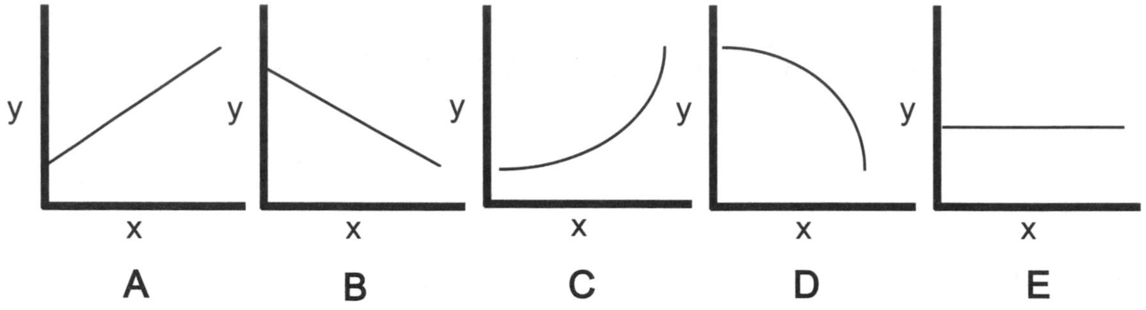

8. Which graph below best describes the variation of ln(K) with temperature when plotting $y = \ln(K)$ and $x = 1/T$, when both ΔH° and ΔS° are negative? Explain your reasoning.

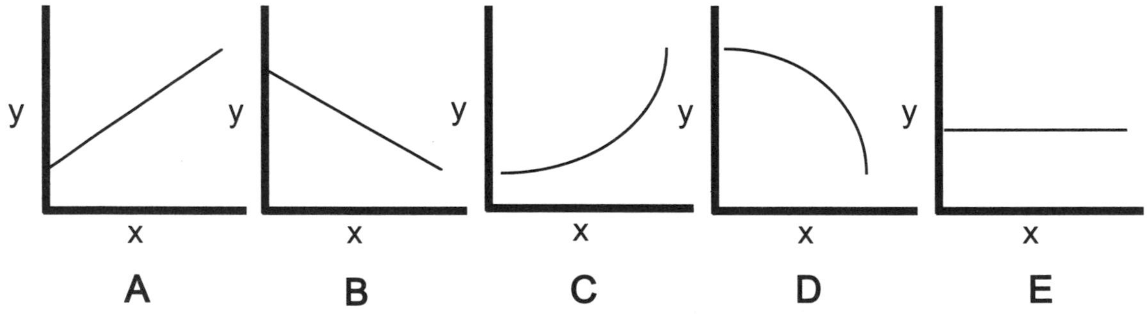

Activity 15-4

Solving Equilibrium Problems: The RICE Table Methodology

Why?

Using a strategy or methodology or following a logical sequence of steps is often helpful in solving problems, especially equilibrium problems. Following the steps of a methodology will help to increase your success as you complete homework assignments, take examinations, and solve problems throughout your life.

Learning Objective

- Master a methodology for solving General Chemistry equilibrium problems

Success Criteria

- Successful application of the RICE table methodology to equilibrium problems
- Accurate solutions to problems involving reactant and product concentrations and equilibrium constants

Prerequisites

- **Activity 05-3:** *Solving Solution Stoichiometery Problems*
- **Activity 15-2:** *The Reaction Quotient and Equilibrium Constant*

Model: The RICE Table Methodology

Initially 1.50 moles of $N_2(g)$ and 3.50 moles of $H_2(g)$ were added to a 1.00 L container at 700 °C. As a result of the reaction,

$$3H_2(g) + N_2(g) \leftrightarrows 2NH_3(g)$$

the equilibrium concentration of $NH_3(g)$ became 0.540 M. What is the value of the equilibrium constant for this reaction at 700 °C?

MODEL: THE RICE TABLE METHODOLOGY (CONTINUED)

The RICE Table Methodology

Steps in the Strategy	Example		
Step R: Write the **Reaction equation**.	$3H_2(g)$ +	$N_2(g)$ $\leftrightarrows$	$2NH_3(g)$
Step I: Write the **Initial amounts** or concentrations of reactants and products.	3.5 mol	1.5 mol	0 mol
Step C: Write the **Change** in the amounts due to the reaction using the stoichiometric coefficients in the reaction equation.	−3x	−x	+2x
Step E: Write the amounts of substances present after the reaction reaches **Equilibrium** or is complete.	3.5 − 3x mol	1.5 − x mol	2x mol
Then solve for unknowns as requested by the problem statement.			

The problem statement gives the concentration of NH_3 at equilibrium, so 2x mol / 1L = 0.540 M, and x = 0.270 mol.

So the equilibrium concentrations are

(3.5 – (3 × 0.270)) mol/L	(1.5 – 0.270) mol/L	(2 × 0.270) mol/L
2.69 M	1.23 M	0.540 M

Substituting into the expression for the equilibrium constant $K = \dfrac{[NH_3]^2}{[H_2]^3[N_2]}$ produces:

$$K = \frac{(0.540)^2}{(2.69)^3(1.23)} = 0.0122$$

KEY QUESTIONS

1. In the RICE table methodology, what do the letters R, I, C, and E represent?

2. Why is writing the balanced reaction equation an important part of the methodology?

3. Are there any steps in the methodology that you feel are not useful? Why?

4. Can you improve the methodology by changing the steps or changing the order of steps? Explain.

5. What insight about solving equilibrium problems did your team gain by examining the methodology?

EXERCISE

1. Initially, 1.0 mole of NO and 1.0 mol of Cl_2 were added to a 1.0 L container. As a result of the reaction,

$$2NO(g) + Cl_2\,(g) \leftrightarrows 2NOCl\,(g),$$

the equilibrium concentration of NOCl became 0.96 M. Using the RICE table methodology, complete the table below and determine the value of the equilibrium constant (K_c) for this reaction.

The RICE Table Methodology

Steps in the Strategy	
Step R: Write the **Reaction equation**.	
Step I: Write the **Initial amounts** or concentrations of reactants and products.	
Step C: Write the **Change** in the amounts due to the reaction using the stoichiometric coefficients in the reaction equation.	
Step E: Write the amounts of substances present after the reaction reaches **Equilibrium** or is complete.	
Then solve for unknowns as requested by the problem statement.	

PROBLEMS

1. If a 10.00 L flask at 500 K is filled with 0.30 mole of hydrogen and 0.30 mole of iodine, what are the equilibrium concentrations of the three gases? The equilibrium constant $K_c = 45.0$. The relevant reaction is

$$H_2(g) + I_2(g) \leftrightarrows 2\ HI(g)$$

2. Nitrogen dioxide is introduced into a flask at a pressure of one atmosphere (1.00 atm). After some time it dimerizes to produce N_2O_4.

   a) What is the equilibrium partial pressure of the N_2O_4 at a temperature where the equilibrium constant for dimerization is $K_p = 3.33$?

   b) Calculate the total pressure in the flask.

   c) Explain why the total pressure is more than, less than, or still equal to one atmosphere.

3. Gaseous carbon dioxide is partially decomposed according to the following equation. An initial pressure of 1.00 atm of CO_2 is placed in a closed container at 2500 K, and 2.1% of the molecules decompose. Determine the equilibrium constant K_p at this temperature.

$$2CO_2(g) \leftrightarrows 2CO(g) + O_2(g)$$

4. A sample of solid ammonium nitrate was placed in an evacuated container and then heated so it decomposed explosively according to the following equation. At equilibrium the total pressure in the container was found to be 4.30 atm at 650 °C. Determine K_p for this reaction.

$$NH_4NO_3(s) \leftrightarrows N_2O(g) + 2H_2O(g)$$

Activity 16-1

The pH Scale and Water Autoionization

Why?

In acid-base reactions in aqueous solution, the active agents are commonly the hydronium ion H_3O^+ or the hydroxide ion OH^-. The range of concentrations of these species is very large, so a logarithmic scale is used to specify these concentrations by a number between 0 and 14. This is the pH scale. You need to understand and be able to use it in order to measure and control acidity and to fully utilize acid-base reactions.

Learning Objectives

- Understand and use the pH scale
- Be able to generalize a logarithmic scale for any quantity, specifically to the hydroxide ion concentration, pOH
- Understand how the pH of pure water is determined by autoionization

Success Criterion

- Correct interconversion of values for the pH, pOH, and hydronium and hydroxide ion concentrations

Prerequisite

- **Activity 04-3:** *Introduction to Acid – Base Reactions*

Model: Logarithms and pH

Properties of Logarithms

$\log(100) = 2$ since $100 = 10^2$

$\log(0.01) = -2$ since $0.01 = 10^{-2}$

$\log(1) = 0$ since $1 = 10^0$

$\log(0.00851) = -2.07$ since $0.00851 = 10^{-2.07}$

$\log(1.122 \times 10^{-14}) = -13.95$ since $1.122 \times 10^{-14} = 10^{-13.95}$

Definition of pH

$pH = -\log[H_3O^+]$

$[H_3O^+] = 0.01\ M = 10^{-2}$ so $pH = 2$

$[H_3O^+] = 0.00851\ M = 10^{-2.07}$ so $pH = 2.07$

$[H_3O^+] = 1.122 \times 10^{-14}\ M = 10^{-13.95}$ so $pH = 13.95$

Key Questions

1. According to the Model, if you are given the logarithm of a number, what do you need to do in order to obtain the number?

2. According to the Model, if you are given the pH of a solution, what do you need to do to obtain the hydronium ion concentration?

3. According to the Model, if you are given the hydronium ion concentration, what do you need to do to obtain the pH of the solution?

4. The pH scale ranges from 0 to 14. What is the range of the hydronium ion concentration covered by this scale?

5. Why is the pH scale used to measure hydronium ion concentrations?

Exercises

1. Using your calculator, check the mathematical calculations shown in the Model to make sure that they are correct, confirming that you know how to interconvert between numbers and their logarithms. Write the results that you got for the following calculations.

   a) $10^2 =$

   b) $\log(0.01) =$

   c) $10^{-2.07} =$

   d) $\log(1.22 \times 10^{-14}) =$

2. Based on the information in the Model, write a definition (using words) for *pH*.

INFORMATION

At 25 °C the hydronium ion concentration of pure water is 10^{-7} M, and the hydroxide ion concentration is also 10^{-7} M due to the autoionization of water.

$$H_2O(aq) + H_2O(aq) \rightleftarrows H_3O^+(aq) + OH^-(aq)$$

Since the concentrations of hydronium ion and hydroxide ion are equal, pure water is said to be neutral. If acids or bases are added to water, the hydronium and hydroxide ion concentrations change. If the hydronium ion concentration increases, the solution becomes acidic. If the hydroxide ion concentration increases, the solution becomes basic.

KEY QUESTIONS

6. What is the pH of a neutral aqueous solution at 25 °C given that the hydronium ion concentration is 1.0×10^{-7} M?

7. Does a pH of 2.7 describe a solution with a higher or lower hydronium ion concentration as compared to a neutral solution? Is such a solution called acidic or basic?

8. Compared to a neutral solution, would you expect a basic solution to have a higher or lower pH and a higher or lower hydronium ion concentration?

9. How could you define a quantity labeled pOH so that it would be analogous to the definition of pH?

INFORMATION

A p-scale can be defined for any quantity as the negative of the logarithm of that quantity:

$$pX = -\log X \text{ so } pOH = -\log[OH^-]$$

Exercises

3. Identify which substance has the highest hydronium ion concentration:

   milk: $[H_3O^+] = 3.2 \times 10^{-7}$ M

   pickle juice: $[H_3O^+] = 2.0 \times 10^{-4}$ M

   oven cleaner: $[H_3O^+] = 3.16 \times 10^{-14}$ M

4. Identify which substance is most acidic and which most basic: milk, pickle juice, or oven cleaner.

5. Identify which substance has the highest pH value: milk, pickle juice, or oven cleaner.

6. Convert the following pH values into hydronium ion concentrations.

pH	**1.0**	**5.3**	**7.0**	**13.3**
Calculated $[H_3O^+]$				

7. Convert the following hydroxide ion concentrations into pOH values first by estimating and then by using your calculator. To estimate, just consider the exponent (power of ten).

$[OH^-]$	**1×10^{-13}**	**2×10^{-9}**	**1×10^{-7}**	**0.2**
Estimated pOH				
Calculated pOH				

8. The values in corresponding columns of the tables in Exercises 6 and 7 were obtained for the same solutions. All the solutions were at 25 °C. For each column in the tables in Exercises 6 and 7, add the pH and calculated pOH values, and put your result in the table below. The first column has been done for you.

(pH + calculated pOH)	**14.0**			

9. For each column in the tables in Exercises 6 and 7, determine the product $[H_3O^+] \times [OH^-]$, and put your result in the table below. The first column has been done for you.

(calculated $[H_3O^+] \times [OH^-]$)	**1.0×10^{-14}**			

10. Write a general statement describing your discovery regarding the sum of the pH and pOH values for an aqueous solution at 25 °C.

11. Write a general statement describing your discovery regarding the product of hydronium and hydroxide ion concentrations for an aqueous solution at 25 °C.

Got It!

1. Milk of Magnesia has a pH of 10.5.

   a) What is the hydronium ion concentration?

   b) Is this commercial product neutral, basic, or acidic?

   c) What is the pOH?

   d) What is the hydroxide ion concentration?

2. As more water is added to an aqueous solution of a strong acid with pH = 3 at 25 °C, the pH (select one):

   a) increases until a constant value of 7 is reached

   b) decreases until a constant value of 0 is reached

   c) doesn't change

   d) increases indefinitely

   e) increases until a constant value of 14 is reached

Activity 16-2

Relative Strengths of Acids

Why?

Many molecules of interest behave as acids or bases. However, not all acids and bases have the same strength. Since the chemical behavior of acids and bases depends upon their strength, knowledge of relative acidity and basicity can be the key to predicting chemical reactivity.

Learning Objective

- Relate previously learned concepts to the relative strengths of acids and bases, including electronegativity, resonance and bond energy, and free energy

Success Criterion

- Predict the relative strengths of acids and bases

Prerequisites

- **Activity 08-2:** *Lewis Model of Electronic Structure*
- **Activity 15-3:** *Free Energy and Chemical Equilibrium*
- **Activity 16-1:** *The pH Scale and Water Autoionization*

Information

The diagram in **Model 1** shows the free energy change for the ionization of an acid. The free energy difference between the products and reactants determines the acid dissociation constant, K_a. When two acids are compared, the acid with the larger K_a is the stronger of the two and will have the less positive change ($G°_{products} - G°_{reactants}$) in free energy ($\Delta G°$). Because the products of the reaction in this model are less stable (i.e., have a higher energy) than the reactant, the degree of ionization is very small and the diagram represents the case of a weak acid. A weak acid is one that does not ionize completely in water. A strong acid ionizes completely in water. Similar diagrams can be drawn for all acids.

Contributed by Frank Fowler and Andisheh Abedini, Stony Brook University
Modified by David Hanson, Stony Brook University

MODEL 1: A REACTION PROFILE SHOWING THE FREE ENERGY CHANGE FOR THE IONIZATION OF A WEAK ACID H-A

Figure 1

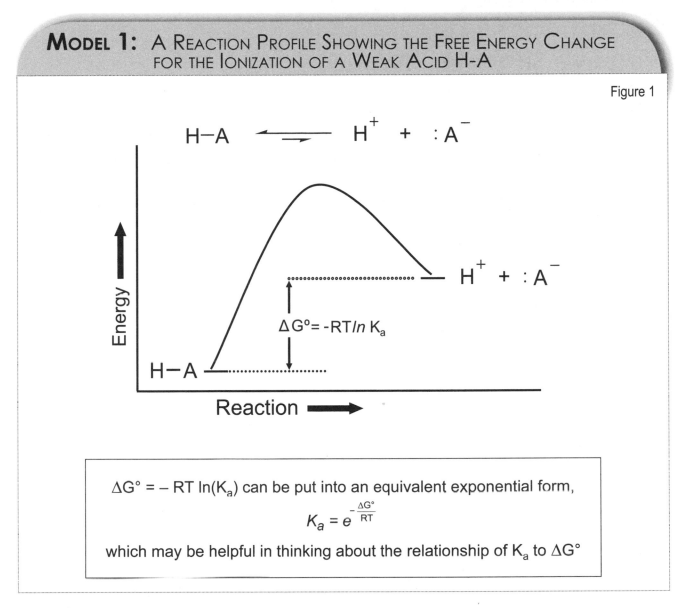

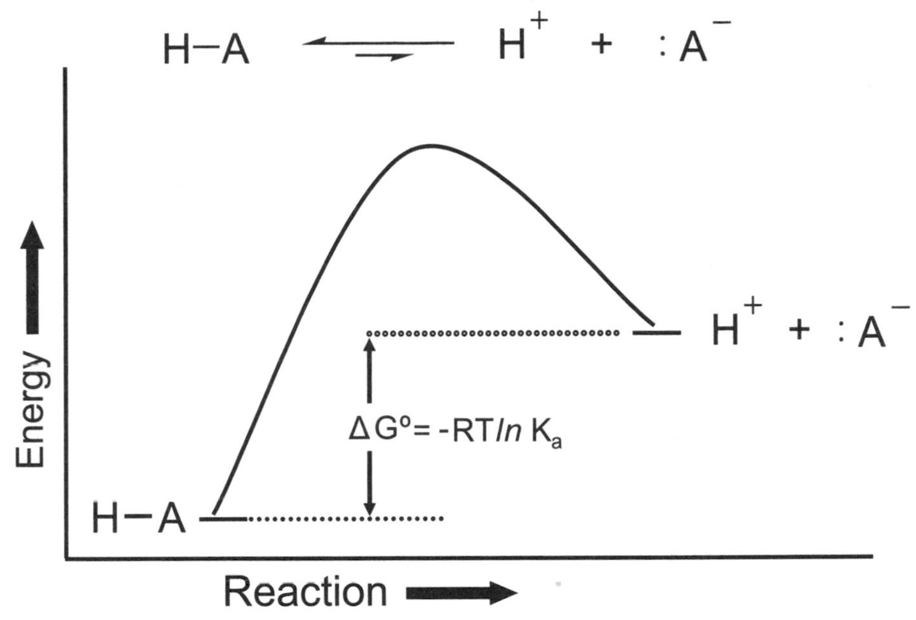

$\Delta G° = -RT \ln(K_a)$ can be put into an equivalent exponential form,

$$K_a = e^{-\frac{\Delta G°}{RT}}$$

which may be helpful in thinking about the relationship of K_a to $\Delta G°$

KEY QUESTIONS

1. According to **Model 1**, which is more stable for the case of a weak acid: the acid HA, or the ionized products H^+ and A^-?

2. What are the two equivalent equations that relate the equilibrium constant to the change in the free energy for the acid ionization in the model?

3. In comparing two acids, will the one with the more positive free energy change have a larger or smaller value for the equilibrium constant, K_a? Provide an explanation for your choice.

Exercises

1. For the weak acid in **Model 1**, write the expression involving molar concentrations for the equilibrium constant, K_a.

2. Draw a sketch below similar to that in Figure 1 comparing the reaction profiles of the weak acid HA in **Model 1** with that of a weak acid, H-B, that is stronger than the weak acid, H-A. Pay particular attention to the difference in energy between the reactants and products of the two reactions.

3. Referring to Exercise 2, predict which conjugate base would be the stronger base: A^- or B^-? Give a reason for your prediction.

4. Which is the stronger acid: H-A ($pK_a = 5.2$), or H-B ($pK_a = 7.4$)? Note: $pK_a = -\log K_a$

5. Referring to Exercise 4, identify the acid with the more positive $\Delta G°$ for ionization: H-A (pK_a = 5.2) or H-B (pK_a = 7.4)?

6. Sketch a reaction profile similar to that in the model but for a strong acid that completely ionizes in water: H-X. Clearly show the energy difference between the reactant and the products.

Model 2: Factors Affecting Acid Strength

As shown in **Model 1**, acid strength is determined by the difference in free energy between the acid and its ionized products. Factors that affect the relative stability of the acid and its conjugate base determine the strength of the acid.

1. **Charge:** Sulfuric acid, H_2SO_4, is a strong acid. It ionizes to produce hydrogen sulfate, HSO_4^-. Hydrogen sulfate, HSO_4^-, is also an acid. It ionizes to produce sulfate, SO_4^{2-}. Hydrogen sulfate is a weaker acid than sulfuric acid because its conjugate base is destabilized by the double negative charge.

2. **Electronegativity:** A conjugate base with a negative charge will be stabilized by electronegative atoms. Consequently, in comparing two bases, the one with more electronegative atoms will be more stable and its conjugate acid will be the stronger of the two.

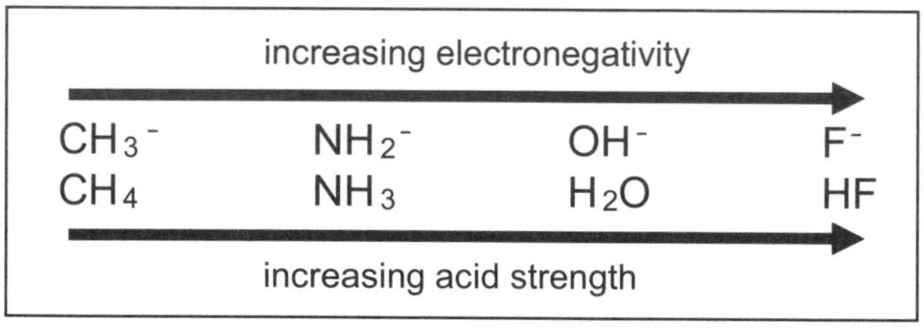

3. **Bond strength:** An acid with stronger bonds will be more stable and therefore will be the weaker acid. This effect can overwhelm the effect of electronegativity as shown in the following figure.

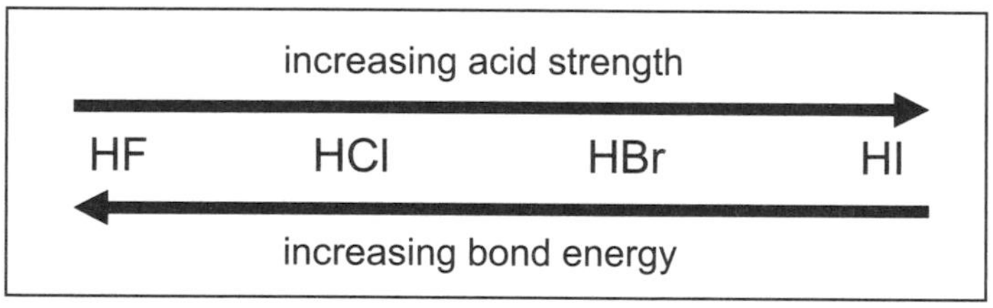

4. **Resonance:** Resonance structures that increase stability are those that delocalize electrons over the molecule and place negative charge on electronegative atoms. For example, HNO_3 is a strong acid and HNO_2 is a weak acid because HNO_3 has more resonance structures that place the negative charge on different oxygen atoms.

Key Questions

4. What are the four factors identified in **Model 2** that affect the strength of an acid?

5. In all factors identified in **Model 2**, what common feature causes them to affect the strength of an acid?

EXERCISE

7. Draw the Lewis resonance structures of NO_3^- and NO_2^- and use them to explain why nitric acid is a strong acid and nitrous acid is a weak acid.

APPLICATIONS

1. Write reaction equations illustrating both H_2O and NH_3 behaving as acids. Explain why H_2O is the stronger acid.

2. Explain why ethanol ($pK_a = 16$) is considerably more acidic than ethane ($pK_a = 50$).

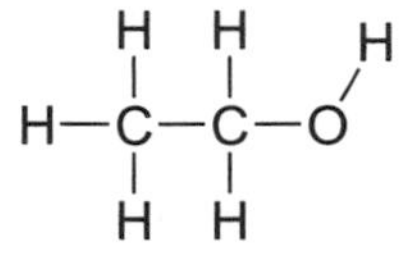

ethanol

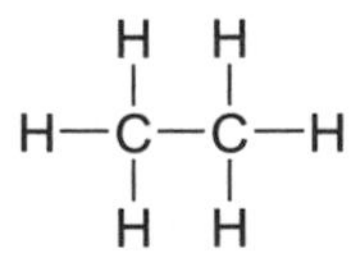

ethane

3. Give one reason why acetic acid is a stronger acid than ethanol.

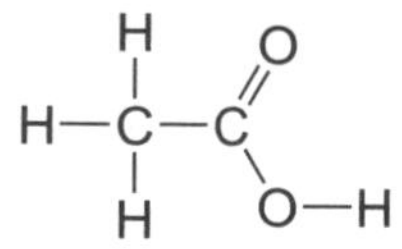

acetic acid

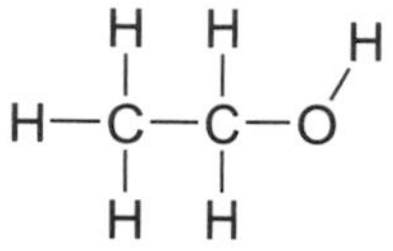

ethanol

4. Normally C-H bonds are not very acidic. An exception is the C-H bond of acetone ($pK_a = 20$). Give an explanation for this observation in terms of a resonance structure.

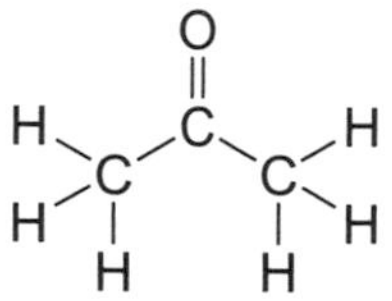

acetone

5. The two carboxylic acid groups of aspartic acid have different acidities with pK_a values of 2.1 and 3.9. Predict which acid group is the more acidic by considering the effect of the $-NH_3^+$ group on the stability of the conjugate bases. Explain your reasoning.

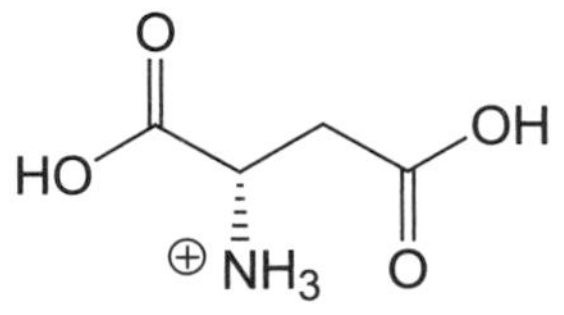

aspartic acid

Activity 16-3

Weak Acid – Base Equilibria

Why?

Weak acids are not as acidic as strong acids at the same concentration because they are not fully ionized. The pH of weak acid and weak base solutions is determined not only by the concentrations of the acids and bases, but also by the extent of ionization. You need to be able to determine the equilibrium concentrations and calculate the pH because such reactions are involved in chemical, biological, and medicinal research and technology.

Success Criteria

- Calculate pH and pOH values from concentrations and equilibrium constant expressions
- Explain and use the relationship between pH and pOH

Prerequisite

- **Activity 16-2:** *Relative Strengths of Acids*

Model 1: Acid Equilibria

Strong Acid Ionization

$$HCl(g) + H_2O(l) \longrightarrow H_3O^+(aq) + Cl^-(aq)$$

Ionization is complete: all of the HCl ionizes.

K_a is very large; $\Delta G°$ is a very large negative number.

$[H_3O^+]$ = moles HCl / volume of solution

$pH = -\log [H_3O^+]$

Weak Acid Ionization

$$HCN(g) + H_2O(l) \longrightarrow H_3O^+(aq) + CN^-(aq)$$

Ionization is not complete: HCN is present in solution.

K_a is a small number, much less than 1; $\Delta G°$ is a large positive number.

$$K_a = \frac{[H_3O^+][CN^-]}{[HCN]} = 6.2 \times 10^{-10}$$

$$[H_3O^+] = \frac{K_a[HCN]}{[CN^-]}$$

$$pH = -\log[H_3O^+] = -\log(K_a) + \log\left(\frac{[CN^-]}{[HCN]}\right)$$

KEY QUESTIONS

1. What is meant by the statements in **Model 1** that ionization is "complete" or "not complete"?

2. If a strong acid is added to water, what determines the pH of the resulting solution as shown by **Model 1**?

3. If a weak acid is added to water, what determines the pH of the resulting solution as shown by **Model 1**?

4. What are five characteristics that distinguish a strong acid from a weak acid?

EXERCISES

1. If a strong acid is added to water, identify what happens to the hydronium ion concentration and the pH.

2. Show how one can use the information in **Model 1** to obtain the final equation for weak-acid ionization:

$$pH = -\log [H_3O^+] = -\log(K_a) + \log\left(\frac{[CN^-]}{[HCN]}\right)$$

3. Rewrite the equation in Exercise 2 using pK_a for any weak acid HA.

4. Calculate the pH of a 0.30 M solution of the strong acid HCl.

5. Calculate the pH of a 1.0 L aqueous solution made from 0.15 mol NaCN and 0.30 mol HCN ($K_a = 6.2 \times 10^{-10}$).

Got It!

1. Which species has the highest concentration in an aqueous hydrocyanic acid solution: HCN, CN^-, or H_3O^+? ($K_a = 6.2 \times 10^{-10}$)

2. How much more base does it take to neutralize a strong acid solution at pH = 2.0 compared to the amount required to neutralize a different strong acid solution at pH = 4.0?

3. How much more base does it take to neutralize a weak acid solution at pH = 2 compared to the amount required to neutralize a different weak acid solution at pH = 4.0?

Model 2: Base Equilibria

Strong Base Ionization

A strong base is typically a metal hydroxide like sodium hydroxide that ionizes or dissociates in water to produce a metal cation and a hydroxide anion.

In water, $NaOH(s) \longrightarrow Na^+(aq) + OH^-(aq)$

Ionization is complete; all of the NaOH ionizes.

$\Delta G°$ is a large negative number.

$[OH^-]$ = moles NaOH/volume of solution

$pOH = -\log [OH^-]$

Weak Base Ionization

A weak base, like ammonia, typically removes a proton from water to produce hydroxide ions.

$$NH_3(g) + H_2O(l) \longrightarrow NH_4^+(aq) + OH^-(aq)$$

Ionization is not complete; NH_3 is present in solution.

K_b is a small number, much less than 1; $\Delta G°$ is a large positive number.

$$K_b = \frac{[NH_4^+][OH^-]}{[NH_3]} = 1.8 \times 10^{-5}$$

$$[OH^-] = \frac{K_b[NH_3]}{[NH_4^+]}$$

Key Questions

5. What is meant by the statements in **Model 2** that ionization of a base is "complete" or "not complete"?

6. If a strong base is added to water, what determines the pOH of the resulting solution as shown by **Model 2**?

7. If a weak base is added to water, what determines the pOH of the resulting solution as shown by **Model 2**?

8. What are five characteristics that distinguish a strong base from a weak base?

Exercises

6. If a strong base is added to water, identify what happens to the hydroxide ion concentration and the pOH.

7. Calculate the pOH of a 0.30 M solution of the strong base NaOH.

8. Using the RICE table methodology developed in Activity 15-4, calculate the pOH of a 0.30 M ammonia solution ($K_b = 1.8 \times 10^{-5}$).

Got It!

4. Ammonium hydroxide (NH_4OH) is commonly advertised as an ingredient in some window cleaners. Rank the following species in order of decreasing concentration in these solutions: NH_4OH, NH_4^+, OH^-, NH_3. Explain your reasoning. (Note that K_b for ammonia = 1.8×10^{-5}.)

5. Identify five similarities between acids and bases with respect to their ionization reactions.

6. Identify five differences between acids and bases with respect to their ionization reactions.

Model 3: K_a and K_b for Conjugate Acid – Base Pairs Are Related

Acid	pK_a	Conjugate Base	pK_b	$pK_a + pK_b$
HCN hydrocyanic acid	9.21	CN^- cyanide ion	4.79	14.00
NH_4^+ ammonium ion	9.25	NH_3 ammonia	4.75	14.00
H_2CO_3 carbonic acid	6.32	HCO_3^- hydrogen carbonate ion	7.68	14.00
H_3O^+ hydronium ion	0.00	H_2O water	14.00	14.00
H_2O water	14.00	OH^- hydroxide ion	0.00	14.00

Note: $pK_a = -\log K_a$ and $K_a = 10^{-pK}$ The pK_a value is easier to write and is often easier to use than the K_a value, which involves writing a number and multiplying it by some power of 10.

Key Questions

9. What relationship between pK_a and pK_b can you identify from the table in **Model 3**?

10. Using your answer to Key Question 9, what relationship between K_a and K_b can you derive using the following property of logarithms: $\log(ab) = \log a + \log b$?

Exercise

9. Each member of your team should pick one acid-base pair from the table in **Model 3**, and write the ionization reaction equation for the acid and the one for the conjugate base. Also write the expressions for K_a and K_b in terms of the molar concentrations. Then combine the reaction equations by adding them, and multiply the K_a and K_b expressions together. The case of HCN and CN^- is given as an example.

(1) acid eq: $HCN + H_2O \rightleftarrows H_3O^+ + CN^-$ $\quad K_a = \dfrac{[H_3O^+][CN^-]}{[HCN]} = 10^{-9.21}$

(2) base eq: $CN^- + H_2O \rightleftarrows HCN + OH^-$ $\quad K_b = \dfrac{[HCN][OH^-]}{[CN^-]} = 10^{-4.79}$

(3) sum: $2\ H_2O \rightleftarrows H_3O^+ + OH^-$ $\quad K_a \times K_b = [H_3O^+][OH^-] = 1.0 \times 10^{-14}$

Note: Equation (3) represents the autoionization of water, which has an equilibrium constant of 1.0×10^{-14} at 25 °C. In this auto or self-ionization reaction, water acts as both an acid (proton donor) and a base (proton acceptor).

Got It!

7. Each member of your team should get similar results for different acid-conjugate-base pairs in Exercise 9. Use these common results to explain why $pK_a + pK_b = 14$ for any conjugate acid-base pair.

Activity 17-1

Buffer Solutions

Why?

Solutions of a weak acid and its conjugate base are called *buffers*. Buffer solutions have specific pH values and resist changes to the pH when they are diluted or when acids or bases are added to them. Such buffering action is important in our blood because cells function only within a narrow pH range. When working with pH-sensitive reactions, you need to introduce the appropriate buffer to control the pH.

Learning Objective

- Understand the nature and properties of buffer solutions

Success Criteria

- Correct identification of buffer systems and their properties
- Accurate calculations of the pH of buffer solutions

Prerequisite

- **Activity 16-3:** *Weak Acid – Base Equilibria*

MODEL: ADDING BASE TO NONBUFFERED AND BUFFERED SOLUTIONS

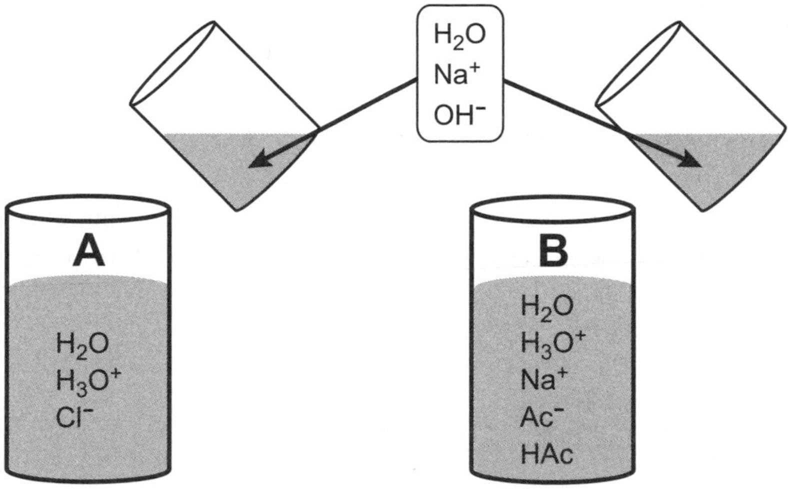

Initially

Beaker A: 100. mL Acid Solution

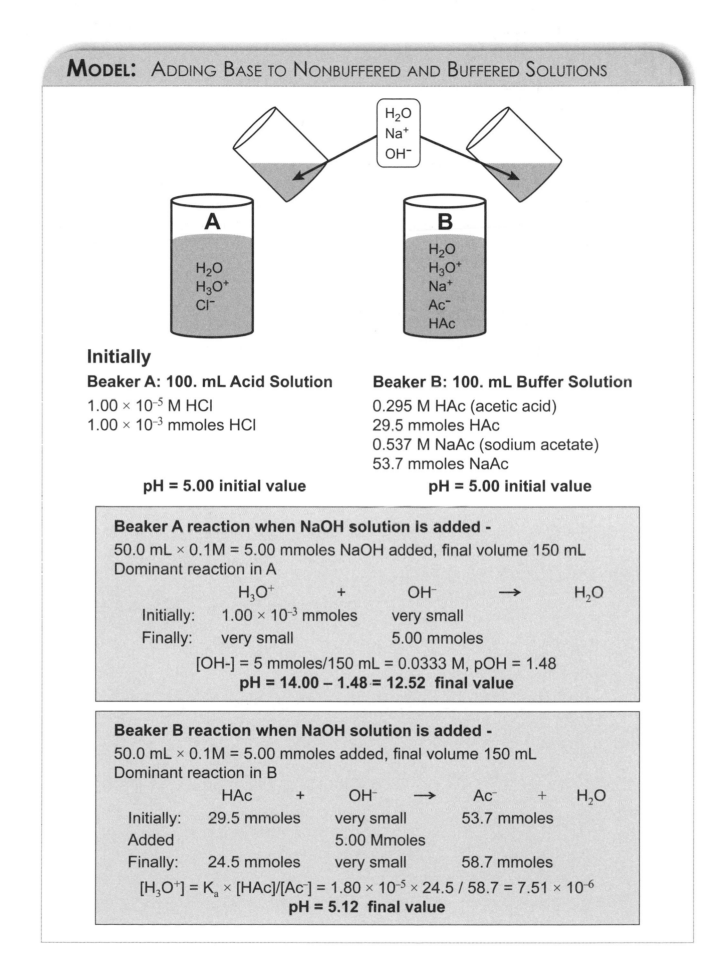

1.00×10^{-5} M HCl
1.00×10^{-3} mmoles HCl

pH = 5.00 initial value

Beaker B: 100. mL Buffer Solution

0.295 M HAc (acetic acid)
29.5 mmoles HAc
0.537 M NaAc (sodium acetate)
53.7 mmoles NaAc

pH = 5.00 initial value

Beaker A reaction when NaOH solution is added -

50.0 mL × 0.1M = 5.00 mmoles NaOH added, final volume 150 mL
Dominant reaction in A

	H_3O^+	+	OH^-	$\longrightarrow$	H_2O
Initially:	1.00×10^{-3} mmoles		very small		
Finally:	very small		5.00 mmoles		

[OH-] = 5 mmoles/150 mL = 0.0333 M, pOH = 1.48
pH = 14.00 – 1.48 = 12.52 final value

Beaker B reaction when NaOH solution is added -

50.0 mL × 0.1M = 5.00 mmoles added, final volume 150 mL
Dominant reaction in B

	HAc	+	OH^-	$\longrightarrow$	Ac^-	+	H_2O
Initially:	29.5 mmoles		very small		53.7 mmoles		
Added			5.00 Mmoles				
Finally:	24.5 mmoles		very small		58.7 mmoles		

$[H_3O^+] = K_a \times [HAc]/[Ac^-] = 1.80 \times 10^{-5} \times 24.5 / 58.7 = 7.51 \times 10^{-6}$
pH = 5.12 final value

Key Questions

1. What are the initial pH values for the solutions in beaker A and beaker B?

2. When 5.00 mmoles of hydroxide are added to beaker A, what does the Model show the pH of the resulting solution to be?

3. When 5.00 mmoles of hydroxide are added to beaker B, what does the Model show the pH of the resulting solution to be?

4. When 5.00 mmoles of hydroxide are added to beaker B, why does the equilibrium amount of hydroxide in the resulting solution remain small, and the pH change only slightly?

5. What equilibrium constant equation is used to calculate the hydronium ion concentration for the solution in beaker B?

6. What parameters are most important in determining the initial and the final hydronium ion concentration of solution B?

7. Suppose you wanted to make a third buffer solution (solution C) with pH = 3. What would you do to make it?

Exercises

1. Identify the essential ingredients of a buffer solution.

2. Calculate the pH of a solution 0.150 M in acetic acid and 0.300 M in sodium acetate.

3. Calculate the ratio $[NH_3]/[NH_4^+]$ needed in an ammonia/ammonium chloride buffer to produce a pH of 8.55.

4. Identify which of the following result in buffer solutions when equal volumes of the two solutions are mixed. ***Note:*** nitric acid is a strong acid and nitrous acid is a weak acid.

a) 0.1 M KNO_3 and 0.1 M HNO_3

b) 0.1 M $NaNO_2$ and 0.1 M HNO_2

c) 0.1 M HCl and 0.1 M NH_3

d) 0.2 M HCl and 0.1 M NH_3

e) 0.1 M HCl and 0.2 M NH_3

Activity 17-2

Acid – Base Titrations

Why?

In a *titration*, an amount of one substance is added to another in the exact stoichiometric proportion so that neither reactant is present in excess. Titrations are widely used for quantitative chemical analyses in industrial quality control, clinical and environmental laboratories, and chemical research. You can use your knowledge of acid-base reactions and equilibria to predict the pH of a solution at various points in the titration and to determine when the stoichiometric point has been reached.

Learning Objectives

- Understand why titration curves have particular shapes
- Identify the information present in a titration curve

Success Criteria

- Accurate calculation of the pH at various points during a titration
- Correct identification of major species in solution during a titration

Prerequisites

- **Activity 05-1:** *Limiting Reactants*
- **Activity 05-3:** *Solving Solution Stoichiometry Problems*
- **Activity 16-3:** *Weak Acid – Base Equilibria*
- **Activity 17-1:** *Buffer Solutions*

Tasks

Analyze the titration of a weak base (ammonia, NH_3) with a strong acid (hydrochloric acid, HCl) by completing Exercises 1 and 2, and Table 1.

Assume that you are titrating 100.0 mL of 0.0500 M ammonia ($K_b = 1.8 \times 10^{-5}$) with 0.100 M hydrochloric acid.

Construct a graph of your results from the data you enter in Table 1.

EXERCISES

1. Before any HCl has been added, the pH is determined by the reaction of ammonia with water.

   Write the reaction of ammonia with water and the corresponding equilibrium constant expression. Determine and write the value for K_b (the equilibrium constant).

2. Write the reaction equation for the reaction of NH_4^+ with water and the corresponding equilibrium constant expression. This equilibrium constant is called K_a, the acid ionization constant. Remember that $K_a = K_w/K_b$. Also write the value for K_a.

INFORMATION

Prior to the stoichiometric point, the pH can be determined from the equations that you wrote in Exercise 1, but you must account for the fact that the HCl that is added converts some NH_3 to NH_4^+. You must also account for the fact that as you titrate, the volume of the solution increases, and you must use this larger volume in calculating concentrations.

At the stoichiometric point, essentially all the NH_3 has been converted to NH_4^+. The strongest acid present in solution at this point is NH_4^+, so the pH can be determined from the acid ionization constant of NH_4^+, as you summarized in Exercise 2.

After the stoichiometric point, the strongest acid in solution is the excess HCl, so the pH is determined by the excess HCl added to the solution.

Calculations Needed to Complete Table 1

1. Before adding HCl:

2. After the addition of 10 ml of HCl:

3. After the addition of 25 ml of HCl:

4. After 49.5 mL of HCl has been added:

5. After 50 ml of HCl has been added:

6. After 50.5 mL of HCl of HCl has been added:

7. After 60 ml of HCl has been added.

8. After 75 ml of HCl has been added.

Calculated Data for the Titration of 100.0 mL 0.0500 M Ammonia with 0.10 M Hydrochloric Acid

Table 1

Situation	Major Species	$[H_3O^+]$	pH
before adding HCl			
10.0 mL HCl added			
25.0 mL HCl added			
49.5 mL HCl added			
50.0 mL HCl added			
50.5 mL HCl added			
60.0 mL HCl added			
75.0 mL HCl added			

Graph of the Titration of 100.0 mL 0.0500 M Ammonia with 0.10 M Hydrochloric Acid

Plot pH on the y-axis and volume HCl added on the x-axis. A high-quality graph has a title, labeled axes, and a scale. Points are plotted with small circles around them to make them easier to see, and a smooth line is drawn through the points.

Key Questions

1. What is the net ionic chemical reaction equation for the reaction of hydrochloric acid with ammonia?

2. Why is the halfway point for this titration considered to be the point at which 25.0 mL of HCl have been added?

3. What are the relative concentrations of NH_4^+ and NH_3 at the halfway point? Provide a qualitative answer, not numbers.

4. Why is the stoichiometric point for this titration considered to be the point at which 50.0 mL of HCl have been added?

5. At the stoichiometric point, what are the concentrations of NH_4^+ and NH_3? Provide a qualitative answer, not numbers.

6. At what point in the titration is the pH equal to the pK_a of the weak acid involved in the titration? Why?

7. What are the major species present in the solution after the stoichiometric point has been reached, e.g., at 75 mL?

8. Why does the pH change so abruptly near the stoichiometric point?

Activity 17-3

Solubility and the Solubility Product

Why?

Many ionic compounds are very soluble in water while others are very insoluble. For example sodium chloride, NaCl, and silver chloride, AgCl, appear to be very similar compounds. Yet sodium chloride is very soluble, and silver chloride is very insoluble in water. This situation results from the balance that is struck between the lattice energy of the ions in the solid, the hydration energy when the ions go into solution, and the change in entropy when the solution is formed. An equilibrium constant, which is called the solubility product constant (K_{sp}), is used to indicate the extent to which any ionic solid is soluble. The solubility product constant enables one to predict how much of any compound will dissolve, how conditions like pH and the presence of other ions affect the solubility, and which ionic compound of several possibilities will precipitate first as conditions change.

Learning Objectives

- Learn how to write the expression for the solubility product constant
- Use a solubility product constant in solving quantitative problems involving solubility

Success Criteria

- Correct quantitative predictions of solubilities given solubility product constants
- Calculate accurate values for solubility product constants given quantitative data on solubilities

Prerequisites

- **Activity 15-2:** *The Reaction Quotient and Equilibrium Constant*
- **Activity 15-3:** *Free Energy and Equilibrium*
- **Activity 15-4:** *Solving Equilibrium Problems*

Information

This activity neglects several complicating factors in order to make the connection between solubility and the solubility product constant as simple as possible. In reality when ionic solids dissociate in water, the ions can react with water. For example, when lead(II) sulfate ($PbSO_4$) dissociates, the sulfate ion (SO_4^{2-}) reacts with water to produce hydrogen sulfate (HSO_4^-). Ion pairs can also be formed. For example, when lead(II) chloride dissociates in water, the solution contains $PbCl^+$ as well as Pb^{2+} and Cl^-. Such effects, which are ignored in this activity, make the compounds more soluble than predicted by values of the solubility product constants.

Model: Solubility and the Solubility Product Constant

Four salts (silver chloride, sodium chloride, lead(II) iodide, and lead(II) sulfate are added to water in four separate beakers until the solutions are saturated and excess solid remains on the bottom of each one.

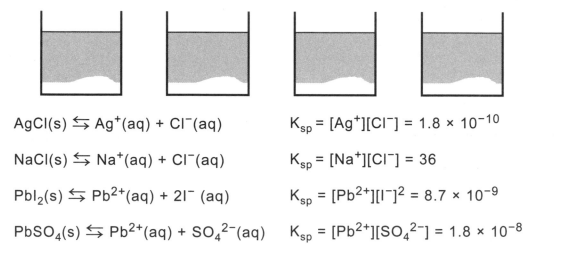

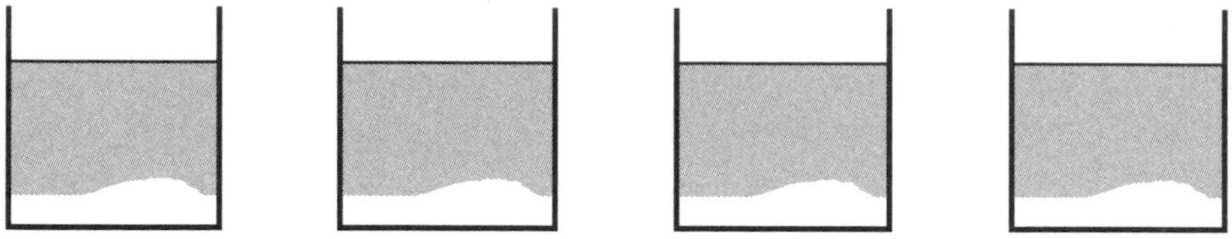

$AgCl(s) \leftrightarrows Ag^+(aq) + Cl^-(aq)$ $\quad K_{sp} = [Ag^+][Cl^-] = 1.8 \times 10^{-10}$

$NaCl(s) \leftrightarrows Na^+(aq) + Cl^-(aq)$ $\quad K_{sp} = [Na^+][Cl^-] = 36$

$PbI_2(s) \leftrightarrows Pb^{2+}(aq) + 2I^-(aq)$ $\quad K_{sp} = [Pb^{2+}][I^-]^2 = 8.7 \times 10^{-9}$

$PbSO_4(s) \leftrightarrows Pb^{2+}(aq) + SO_4^{2-}(aq)$ $\quad K_{sp} = [Pb^{2+}][SO_4^{2-}] = 1.8 \times 10^{-8}$

Key Questions

1. How is the form of the solubility product constant similar to, and how does it differ from, the standard form of an equilibrium constant for a chemical reaction equation?

2. Is silver chloride or sodium chloride more soluble in water? Explain how specific information in the Model supports your conclusion.

3. In general, in comparing solubility product constants to predict solubilities as you did in Key Question 2, what additional information needs to be taken into account in addition to the K_{sp} values? Carefully consider the situation for lead(II) iodide and lead(II) sulfate in the Model.

4. Suppose chloride ions could be removed selectively from the AgCl solution in the Model. What would happen to the concentration of the silver ion as a result? Explain.

5. Suppose additional chloride ions could be added to the AgCl solution in the Model. What would happen to the concentration of the silver ions as a result? Explain.

Exercises

1. What is the solubility of barium sulfate ($K_{sp} = 1.1 \times 10^{-10}$) in moles/L and in g/L?

2. A saturated solution of MgF_2 contains 1.2×10^{-3} M Mg^{2+}. What is the value of the solubility product constant for magnesium fluoride?

3. How much lead(II) chloride will dissolve in a solution that is 0.1 M in sodium chloride? The K_{sp} for lead (II) chloride is 1.7×10^{-5}.

PROBLEMS

1. Lead pipes were once used to transport drinking water and are still found in some older homes and city water systems. Suppose the lead concentration and pH of the water were determined by lead (II) hydroxide ($K_{sp} = 2.8 \times 10^{-16}$) dissolving to produce a saturated solution. What would the lead concentration (M) and pH of this water be?

2. Although water is normally slightly acidic due to the presence of dissolved carbon dioxide and metal ions, some municipal water utilities deliver water at a relatively high pH to limit the amount of lead that can dissolve from any lead pipes remaining in the system. Based on the equilibrium constant expression for lead (II) hydroxide, how much lower is the maximum possible lead(II) ion concentration at pH 8.0 than at pH 6.0? To answer this question, calculate the ratio: (concentration at pH 8.0) / (concentration at pH 6.0).

3. In a mining operation it necessary to separate copper (I) ions from lead (II) ions. The concentration of each of these ions in solution is 1.0×10^{-3} M. The strategy is to gradually add sodium iodide to precipitate one iodide salt before the other.

   a) Which salt precipitates first, CuI ($K_{sp} = 5.1 \times 10^{-12}$) or PbI_2 ($K_{sp} = 8.7 \times 10^{-9}$)?

   b) What is the iodide concentration necessary to begin the precipitation of each salt?

   c) What is the remaining concentration of the ion that began precipitating first at the point where

d) What percent of the original copper concentration remains in solution?

4. Although silver chloride is very insoluble in water, it can be dissolved in a solution of aqueous ammonia due to the following series of reactions leading to the formation of the diamminesilver complex ion.

$$AgCl(s) \leftrightarrows Ag^+(aq) + Cl^-(aq) \qquad K_{sp} = 1.8 \times 10^{-10}$$

$$Ag^+(aq) + NH_3(aq) \leftrightarrows Ag(NH_3)^+ (aq) \qquad K_{f1} = 2.1 \times 10^3$$

$$Ag(NH_3)^+ (aq) + NH_3(aq) \leftrightarrows Ag(NH_3)_2^+ (aq) \qquad K_{f2} = 8.2 \times 10^3$$

a) Determine the value of the equilibrium constant for the overall reaction:

$$AgCl(s) + 2\,NH_3(aq) \leftrightarrows Ag(NH_3)_2^+(aq) + 2Cl^-(aq)$$

b) Determine the amount (moles) of silver chloride that will dissolve in 1.0 L of a 10.0 M ammonia solution.

Activity 18-1

Voltaic Cells

Why?

Voltaic electrochemical cells are also known as *galvanic cells* or batteries. They are used to store energy and provide electrical power on demand. Because voltaic cells are ubiquitous, it is important to understand how one works and the factors that determine its voltage.

Learning Objective

- Understand how a voltaic cell produces electricity by a chemical reaction

Success Criteria

- Identify the cell components and sites of oxidation and reduction
- Calculate cell potentials and free energy changes for cell reactions

Prerequisites

- **Activity 04-4:** *Electron Transfer Reactions*
- **Activity 15-3:** *Free Energy and Chemical Equilibrium*

Information

A redox reaction can be separated into two half-reactions, an oxidation reaction and a reduction reaction. In electrochemical cells, each reaction occurs in a different compartment, and electrons are transferred through a wire from one compartment to the other.

Reduction is the chemical reaction that reduces the oxidation number of a chemical species by making the oxidation number less positive.

Oxidation is the chemical reaction that increases the oxidation number of a chemical species by making the oxidation number more positive.

The *standard reduction potential*, E°_{red}, is a measure of the strength of a reduction reaction's ability to attract electrons under standard conditions (pressure = 1 atm, temperature = 25 °C, concentrations = 1 M). The more positive the reduction potential, the stronger the attraction for electrons. Tables of standard reduction potentials can be found in chemistry reference and text books.

In a voltaic electrochemical cell, the half-reaction with the more positive reduction potential attracts electrons more strongly and proceeds as a reduction reaction. The other half-reaction proceeds as oxidation. Since the oxidation half-reaction is just the reverse of the corresponding reduction half-reaction, the standard oxidation potential, E°_{ox} equals $-E^\circ_{red}$.

A *cathode* is the site of reduction. An *anode* is the site of oxidation.

Note the following relationship: 1 *Joule* (J) of electrical energy is released by 1 *Coulomb* (C) of charge flowing through a potential difference of 1 *volt* (V).

Model 1: The Zn/Cu Voltaic Cell

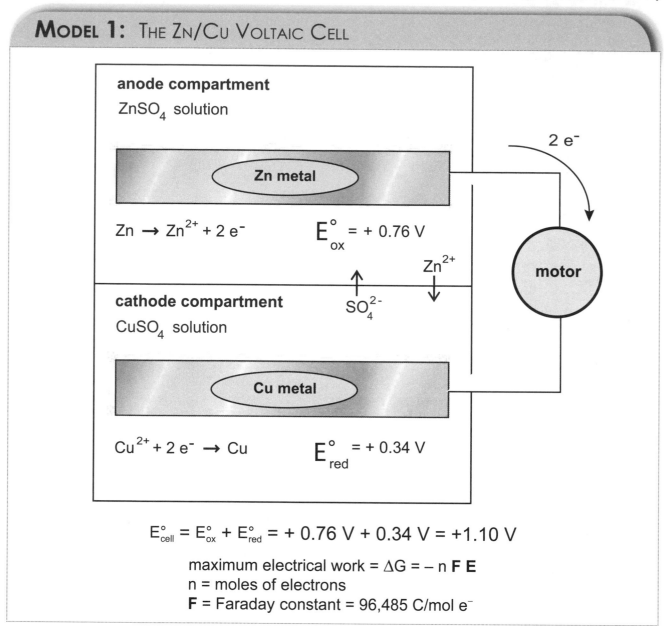

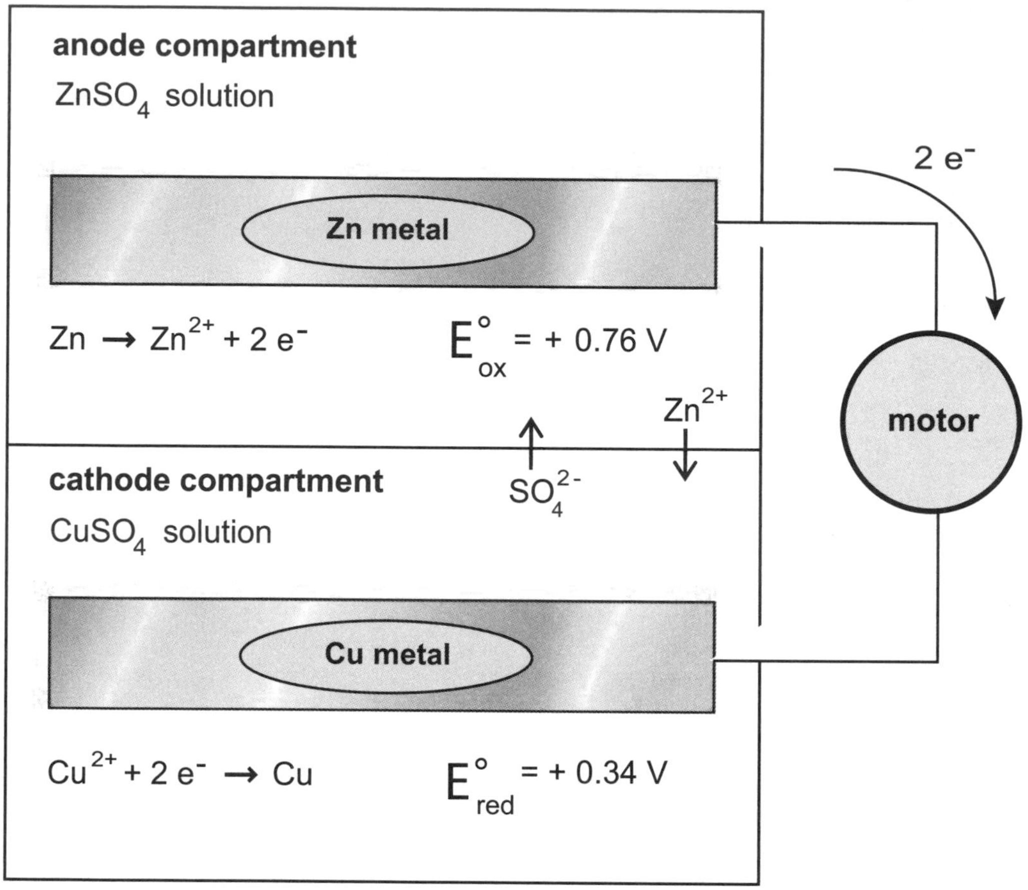

$$E^\circ_{cell} = E^\circ_{ox} + E^\circ_{red} = +\,0.76\text{ V} + 0.34\text{ V} = +1.10\text{ V}$$

maximum electrical work = $\Delta G = -\,n\,\mathbf{F}\,\mathbf{E}$
n = moles of electrons
F = Faraday constant = 96,485 C/mol e^-

Key Questions

1. In the model, which reaction takes place at the anode: oxidation or reduction? Identify the species that is oxidized or reduced.

2. Which reaction takes place at the cathode: oxidation or reduction? Identify the species that is oxidized or reduced.

3. Is the Zn metal an oxidizing agent or a reducing agent?

4. Which way do electrons flow through the wire, from the anode to the cathode or from the cathode to the anode? Explain why you think this would happen.

5. Since electrons flow from the negative electrode to the positive electrode, which is positive: the Zn metal, or the Cu metal?

6. The model shows that sulfate ions flow into the anode compartment, and zinc ions flow into the cathode compartment. Why do these ions flow between these two compartments of the cell?

MODEL 2: THE STANDARD CELL POTENTIAL AND STANDARD FREE ENERGY CHANGE

$$E^\circ_{cell} = E^\circ_{ox} + E^\circ_{red} = +\,0.76\text{ V} + 0.34\text{ V} = +1.10\text{ V}$$

$$\text{maximum electrical work} = \Delta G = -\,n\,\mathbf{F}\,\mathbf{E}$$

$$\Delta G = \text{free energy change}$$

F = Faraday constant, which is the magnitude of the charge associated with a mole of electrons = 96,485 C/mol e^-

KEY QUESTIONS

7. According to **Model 2**, how is the standard cell potential calculated from the standard half-cell potentials?

8. The standard reduction potential for $Zn^{2+} + 2e^- \longrightarrow Zn$ is $E^\circ_{red} = -0.76$ V. Why is the value + 0.76 V used in **Model 2** to calculate the standard cell potential?

9. Which is the stronger oxidizing agent: Zn^{2+}or Cu^{2+}? Explain how you can tell in terms of the standard reduction potentials.

10. What determines whether a particular half-reaction in a voltaic cell runs as an oxidation reaction or a reduction reaction?

11. According to **Model 2**, how are the free energy change and the cell potential related?

12. Is it essential for the cell reaction in a voltaic cell to be spontaneous? Explain.

13. For the cell reaction to be spontaneous, what must be true of the free energy change and the cell potential?

14. Why might batteries for sale at Radio Shack have different voltages?

EXERCISES

The following two half-reactions are used in a voltaic cell. A strip of silver metal is placed in a 1 M solution of silver nitrate, and a strip of nickel metal is placed in a 1 M solution of nickel nitrate. The metal strips are connected by a wire, and the solutions are connected by a salt bridge. Use this information in Exercises 1 through 7.

$$Ag^+ + e^- \longrightarrow Ag \qquad E^\circ_{red} = +0.80\ V$$

$$Ni^{2+} + 2e^- \longrightarrow Ni \qquad E^\circ_{red} = -0.25\ V$$

1. Sketch this voltaic cell (similar to that shown in **Model 1**).

2. Write and label the reactions occurring at the anode and at the cathode as "oxidation" or "reduction."

   anode:

   cathode:

3. Write the overall cell reaction.

4. Identify the species being oxidized and the species being reduced.

5. Identify the oxidizing agent and the reducing agent.

6. Calculate the standard cell potential E°_{cell} and the standard free energy change ΔG° for the cell reaction.

7. Identify which of the following does not take place in the Ag/Ni cell.

   a) Electrons flow from the nickel electrode to the silver electrode.

   b) The silver electrode increases in mass as the cell operates.

   c) Ag^+ ions move through the salt bridge to the nickel half-cell.

   d) Negative ions move through the salt bridge from the silver half-cell toward the nickel half-cell.

   e) Ni^{2+} ions move from the nickel half-cell toward the silver half-cell.

8. Using the table of standard reduction potentials in your textbook, determine the standard cell potential and standard free energy change for the voltaic cell with the following overall reaction.

$$Cu(s) + 2\,Ag^+(aq) \longrightarrow Cu^{2+}(aq) + 2\,Ag(s)$$

9. Using the table of standard cell potentials in your textbook, identify which of the following species is the strongest oxidizing agent and which is the weakest oxidizing agent.

Ni^{2+} Fe^{3+} I_2 O_2 $Cr_2O_7^{2-}$ Cl_2

10. Using the table of standard cell potentials in your textbook, identify which of the following species is the strongest reducing agent and which is the weakest reducing agent.

Li Na Al Sn Cu I^- Br^-

11. To reduce pollution when coal is burned, it is important to remove mercury and sulfur compounds from the effluent. Identify which, if any, of the following redox reactions are spontaneous under standard conditions and might therefore be used for this purpose.

a) $O_2 + 2\ Hg \longrightarrow 2\ HgO$

b) $Cu + SO_2 \longrightarrow CuS + O_2$

c) $Fe + SO_2 \longrightarrow FeS + O_2$

Activity 18-2

Electrolytic Cells

Why?

An electrolytic cell uses electrical energy to produce chemical change. In working with electrolytic cells, you need to be able to identify the voltage and current necessary to operate the cell, as well as the chemicals needed in the anode and cathode compartments. Electrolysis is used to synthesize compounds, charge the battery in your car, plate one metal over another (e.g., silverplate dinnerware), and produce aluminum.

Learning Objective

- Understand how an electrolytic cell uses electricity to cause chemical reactions

Success Criteria

- Identify the components of an electrolytic cell
- Identify the chemical reactions that occur at each electrode
- Determine the amount of material produced or consumed and the electrical current used

Prerequisite

- **Activity 18-1:** *Voltaic Cells*

Information

Electrical current is measured in amperes (A) where 1 A = 1 C/s.

1 C (coulomb) is the charge on 6.2415×10^{18} electrons.

1 Faraday is the charge on 1 mole of electrons. 1 **F** = 96,485 C/mol e^-

Since an electrolytic cell is not spontaneous, the standard cell potential for an electrolytic cell that is calculated from the standard half-cell potentials is negative. This cell potential is the minimum voltage that must be applied to drive the reactions in the direction that is not spontaneous.

Model: The Zn/Cu Electrolytic Cell

Figure 1

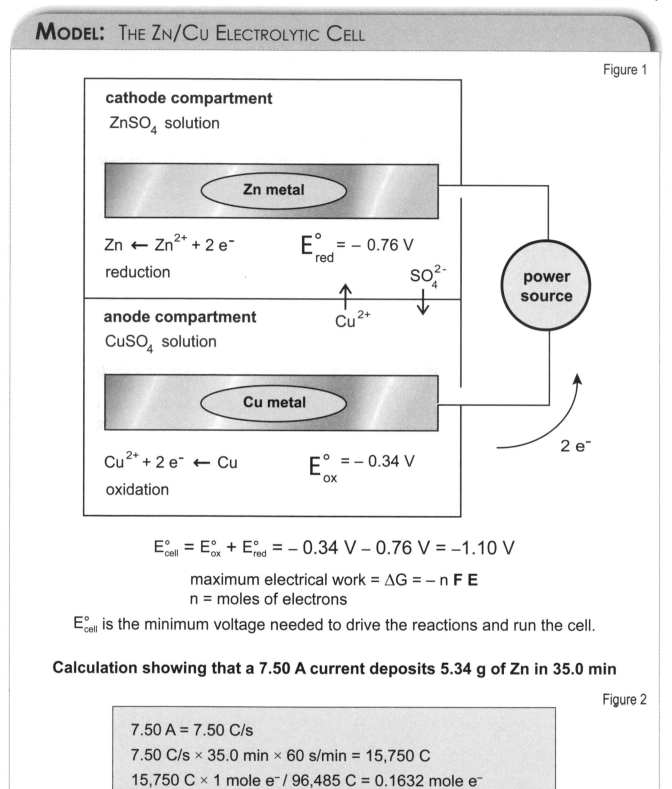

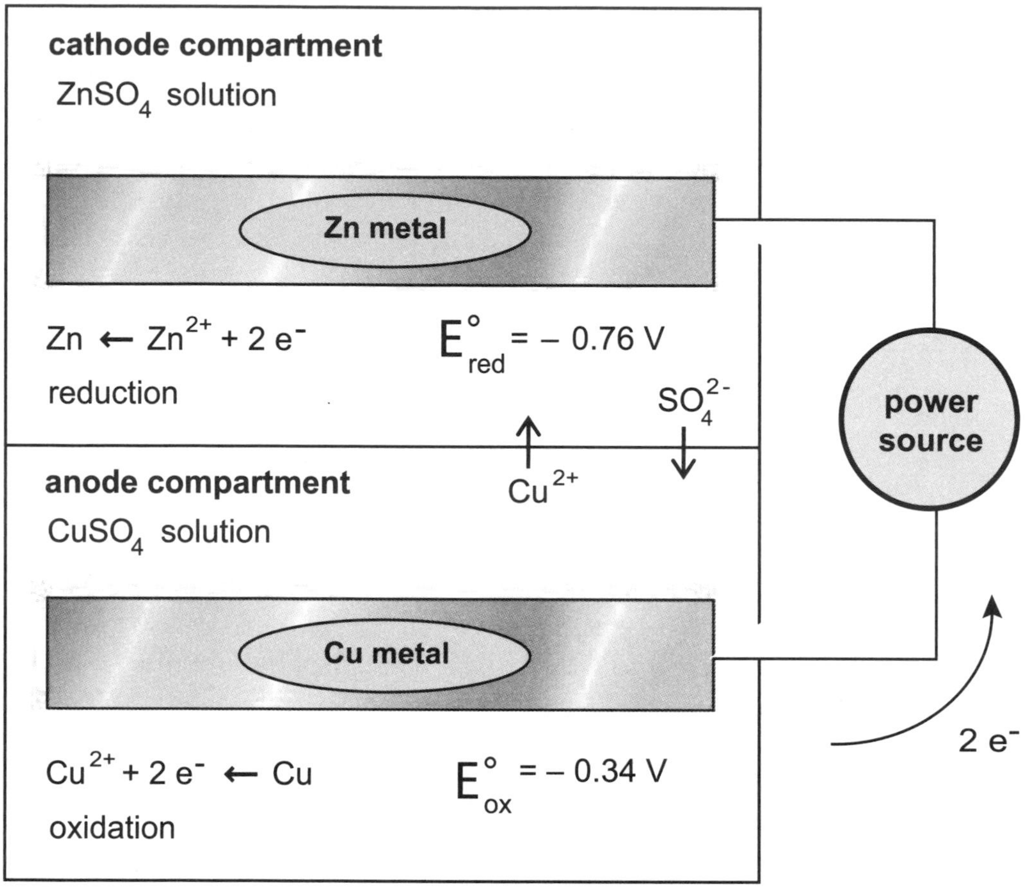

$$E^\circ_{cell} = E^\circ_{ox} + E^\circ_{red} = -0.34\text{ V} - 0.76\text{ V} = -1.10\text{ V}$$

maximum electrical work = $\Delta G = -n\,\mathbf{F}\,\mathbf{E}$

n = moles of electrons

E°_{cell} is the minimum voltage needed to drive the reactions and run the cell.

Calculation showing that a 7.50 A current deposits 5.34 g of Zn in 35.0 min

Figure 2

7.50 A = 7.50 C/s

7.50 C/s × 35.0 min × 60 s/min = 15,750 C

15,750 C × 1 mole e^- / 96,485 C = 0.1632 mole e^-

0.1632 mole e^- × 1 mole Zn/2 moles e^- = 0.08162 mole Zn

0.08162 mole Zn × 65.38 g/mole = 5.34 g of Zn

KEY QUESTIONS

1. What are five or more important differences between the Zn/Cu voltaic cell and the Zn/Cu electrolytic cell illustrated in the models in this activity and the previous one?

2. According to the Model, what is the minimum voltage that is needed to drive the reactions and run the cell under standard conditions? How is this value determined from standard reduction potentials for the half-reactions?

3. Why do the Cu/Zn voltaic and electrolytic cells have cell potentials with the same magnitude (1.10 V) but different signs (one is positive and the other negative)?

4. Is the free energy change for the electrolytic cell reaction positive or negative? Explain why in terms of the cell potential and whether or not the reaction is spontaneous.

5. A calculation that is part of the Model shows that an electrical current of 7.50 A deposits 5.34 g of Zn in 35.0 min. By examining this calculation:

   a) How many moles of electrons flowed through the cell in 35.0 minutes using 7.50 A of current?

b) How many moles of zinc were plated onto the cathode in 35 minutes using 7.50 A of current?

c) Why is the ratio (moles of electrons)/(moles of Zn) = 2?

6. Name an item you use that might have been produced by electrolysis.

Exercises

1. The electrolysis of water involves the following two half-reactions.

$$2\ H_2O + 2e^- \longrightarrow H_2 + 2\ OH^- \qquad\qquad 2\ H_2O \longrightarrow O_2 + 4\ H^+ + 4e^-$$

a) Write the overall balanced equation for the cell reaction. ***Note:*** electrons cannot appear in the overall reaction equation.

b) Identify the species being oxidized and the species being reduced.

c) Calculate the standard cell potential using information in a table of standard reduction potentials.

d) Calculate the change in free energy under standard conditions.

2. Determine how long it will take to plate out 1.30 kg of aluminum in an electrolytic cell with a current of 125 A.

Activity 19-1

Coordination Compounds: An Introduction

Why?

Transition metals form unique compounds called *coordination compounds*. You need to know the characteristics of these compounds because they are vital to biological process and are widely used in industry, technology, and medicine. Some examples of applications include jewelry, steel, paints, photographic films, magnetic tapes, anticancer drugs, and contrast agents in magnetic resonance imaging (MRI).

Success Criteria

- Identify the components of coordination compounds and their structures
- Identify the geometry of coordination compounds
- Explain the bonding in coordination compounds

Prerequisites

- **Activity 04-4:** *Electron Transfer Reactions*
- **Activity 07-4:** *Multi-electron Atoms, the Aufbau Principle, and the Periodic Table*
- **Activity 08-2:** *Lewis Model of Electronic Structure*
- **Activity 09-1:** *Valence Shell Electron Pair Repulsion Model*
- **Activity 09-3:** *Hybridization of Atomic Orbitals*

Information

A Lewis base is an electron pair donor. A Lewis acid is an electron pair acceptor. In many cases a chemical bond is formed by the reaction between a Lewis acid and a Lewis base. For example, BF_3 is electron deficient because the octet rule is not satisfied for boron, and NH_3 has a nonbonding pair of electrons that can be used to form a bond with BF_3 to yield F_3B-NH_3. In this example, ammonia is the Lewis base, and boron trifluoride is the Lewis acid.

Just as sp^3 hybridization of atomic orbitals is used to describe a tetrahedral electron geometry around a central atom, sp^3d hybridization is needed for a trigonal bipyramidal geometry and sp^3d^2 hybridization is needed for an octahedral geometry.

Contributed by Gulnar Rawji, Southwestern University

Model: Coordination Compounds

Abbreviations:

CN = coordination number
Ox = oxidation state of central metal
Charge = charge on the complex ion

Coordination compound	Complex ion	Geometry	Ligand(s)	Counter ion	CN	Ox	Charge
$[Co(NH_3)_6]Cl_3$	$[Co(NH_3)_6]^{3+}$	Octahedral	NH_3	Cl^-	6	+3	+3
$K_3[Fe(CN)_6]$	$[Fe(CN)_6]^{3-}$	Octahedral	CN^-	K^+	6	+3	–3
$Na_2[HgI_4]$	$[HgI_4]^{2-}$	Tetrahedral	I^-	Na^+	4	+2	–2
$[Pt(NH_3)_2Cl_2]$	None	Square planar	NH_3, Cl^-	None	4	+2	0
$K_2[PtCl_6]$	$[PtCl_6]^{2-}$	Octahedral	Cl^-	K^+	6	+4	–2

Figure 1 $[Co(NH_3)_6]^{3+}$ ion

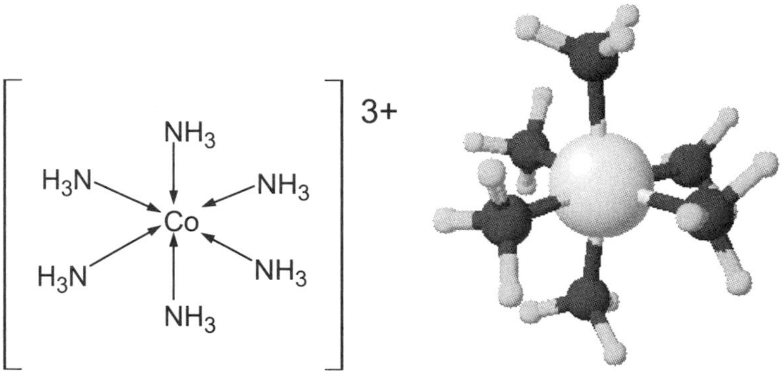

Key questions

1. What are the components of a coordination compound that are present in all such compounds?

2. What additional component may be present in some coordination compounds?

3. Why are the square brackets included in the chemical formula for a coordination compound?

4. What information is provided by the coordination number?

5. When a coordination compound dissolves in water, it dissociates into the complex ion and the counter ions. What are the dissociation products of each coordination compound in the table?

6. How can the charge on the complex ion be calculated from other information in the Model?

7. How can the oxidation number of the transition metal be determined from other information in the Model?

8. In looking at the Lewis electron structures of the ligands, what do they all have in common? How does this characteristic contribute to the formation of a coordination compound in terms of the Lewis acid-base model?

INFORMATION

- A *ligand* is a molecular or an anionic species that has atoms with nonbonding electrons available for donation to the metal.
- A *coordination compound* or *complex* is a Lewis acid-base complex that contains a transition metal ion and several ligands covalently bonded to it. Such a complex can be cationic, anionic, or neutral.
- A *coordinate covalent bond* is a bond in which the shared electrons are contributed by one of the two bonded atoms. Once formed, the resulting bond is no different from a covalent bond in which both atoms of a bond contribute electrons. Arrows in Figure 1 indicate that the N atom in NH_3 uses a lone pair to bond to the metal.
- The *Coordination number* is the number of atoms bonded to the transition metal in a coordination complex.
- *Counter ions* in a coordination compound serve to neutralize the charge on a complex ion.

- A *polydentate ligand* contains more than one atom with nonbonding electrons available for donation. Similarly, a ligand with only one atom that can donate electrons is referred to as *monodentate*; one with two such atoms is a *bidentate ligand*, etc. Examples are shown in Figure 2.
- A *chelating agent* is a polydentate ligand which, when bonded to the metal, results in a complex ion with one or more rings as shown in Figure 3.

Figure 2 Mono- and bidentate ligands

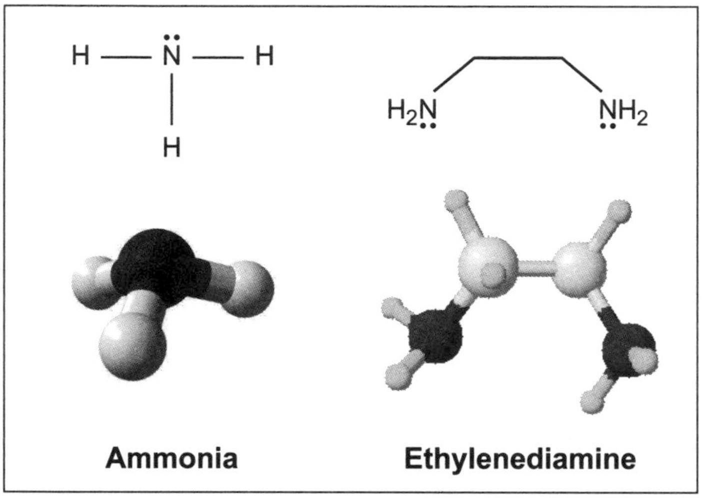

Figure 3 Polydentate ligand and its complex

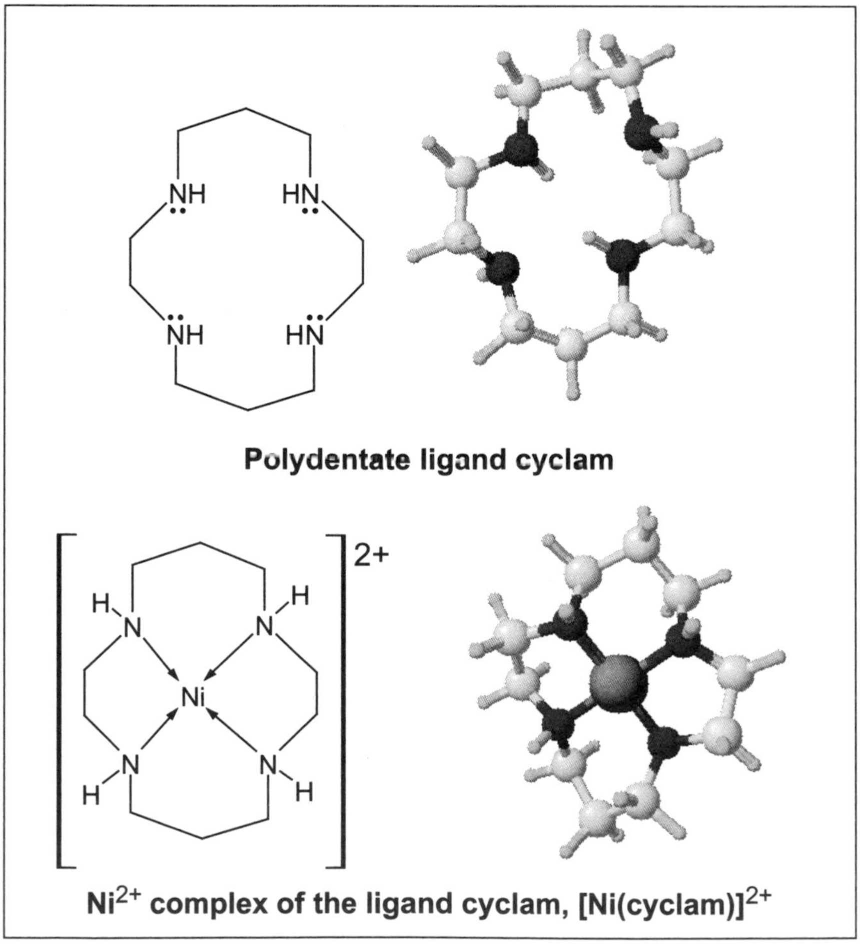

EXERCISES

1. Sketch $[Co(NH_3)_6]Cl_3$ showing its octahedral geometry, and the location of the ligands and counter ions. Label the bonds appropriately as covalent, coordinate covalent, and ionic.

2. Sketch the complex ion only ($[Co(NH_3)_6]^{3+}$) showing the sp^3-orbitals of the ligands overlapping with the sp^3d^2 orbitals of the transition metal ion.

3. Sketch the tetrahedral and square planar geometries of complex ions with coordination number 4.

4. Illustrate the difference between the structures of complex ions containing Fe^{3+} bonded to the polydentate ligand, ethylenediamine (Figure 2), and to the monodentate ligand, NH_3. In both cases the coordination number of Fe is 6. Remember to include the charge on the complex ion.

5. Using your experience with VSEPR theory, construct a table listing possible coordination numbers (2–6) and resulting geometries for complex ions.

Activity 19-2

Coordination Compounds: Magnetism & Color

Why?

Coordination compounds exhibit characteristic magnetic and color properties. These properties arise from the energy splitting of the transition metal's d-orbitals, caused by the presence of ligands in coordination compounds. You need to recognize which d-orbitals are occupied by electrons, and how this occupation determines the magnetic properties and color of the compound. These properties are often used as a probe of the bonding in coordination compounds and to tailor-make compounds with useful magnetic and color characteristics.

Learning Objectives

- Understand the ligand field model (aka: crystal field model)
- Understand magnetism and color in coordination compounds

Success Criteria

- Draw d-orbital energy level diagrams for octahedral coordination complexes
- Identify whether the transition metals are low spin or high spin
- Identify whether the coordination compound is diamagnetic or paramagnetic
- Relate the magnitude of the ligand field splitting to the color of the compound

Prerequisites

- **Activity 07-1:** *Electromagnetic Radiation*
- **Activity 19-1:** *Coordination Compounds: An Introduction*

Contributed by Gulnar Rawji, Southwestern University

Model 1: Ligand Field Model for a Coordination Complex with Octahedral Geometry

According to the ligand field model, the splitting of the energies of the d-orbitals is a function of the relative positions of the ligands and the individual d-orbitals.

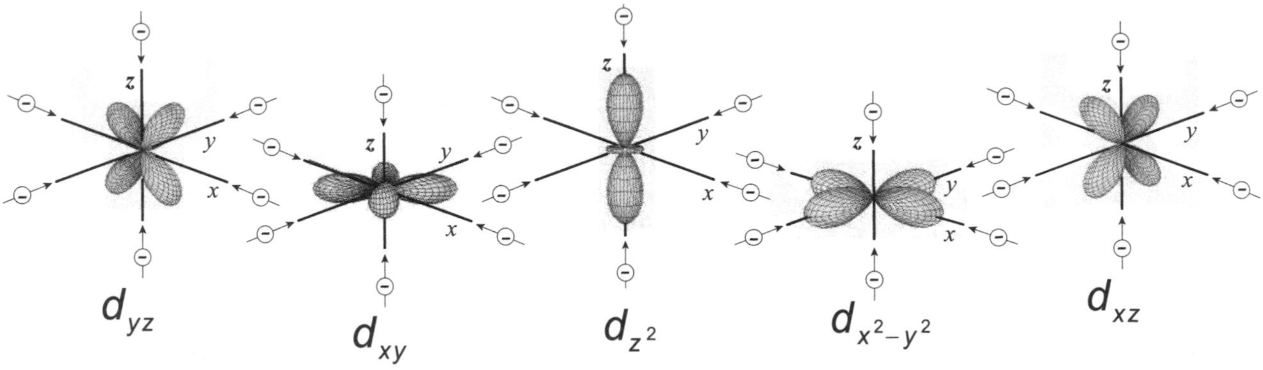

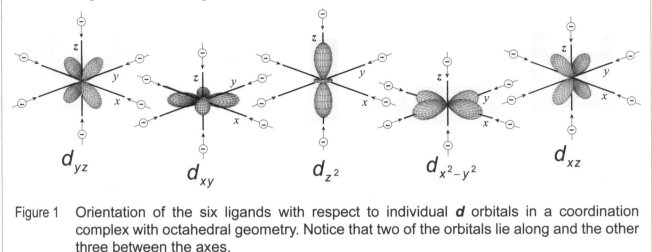

Figure 1 Orientation of the six ligands with respect to individual ***d*** orbitals in a coordination complex with octahedral geometry. Notice that two of the orbitals lie along and the other three between the axes.

Key Questions

1. Which d-orbitals lie along the x, y, or z-axes, and which lie between any two axes?

2. Which orbitals experience a direct head-on interaction with the ligands that also lie along the axes?

3. Which orbitals do not experience a direct head-on interaction with the ligands?

Exercises

1. Based on the nature of the interactions that you identified in Key Questions 2 and 3, divide the five d-orbitals into two sets. Explain the reasoning for your choices.

Set 1	Set 2

2. Consider the electrons. In which set of orbitals will the stronger interactions with the ligands be found: Set 1 or Set 2? Explain the reason for your choice.

3. The diagram below shows all five d-orbitals in a "free metal ion" (i.e., in the absence of ligands) at the same energy level:

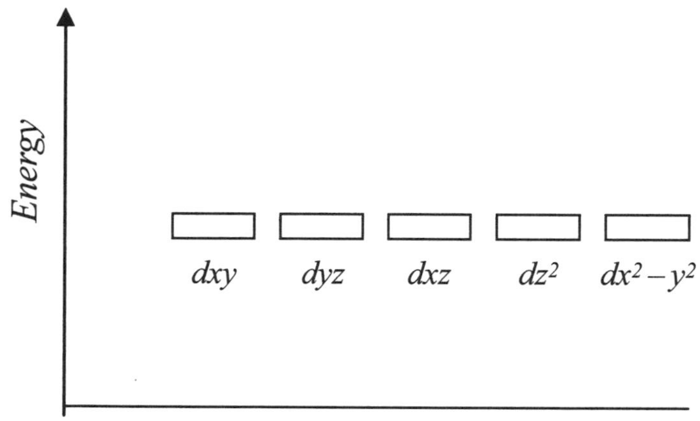

In a compound with octahedral geometry, the energies of the two sets of orbitals depend on the extent of interactions with the ligands. The set experiencing stronger interactions will be higher in energy because the electron-electron interactions are repulsive. Draw a diagram below to show the relative energies of the two sets that you identified and characterized in Exercises 1 and 2.

Model 3: Colors of Coordination Compounds & Magnitude of Δ_O

Primary Colors and the Color Wheel

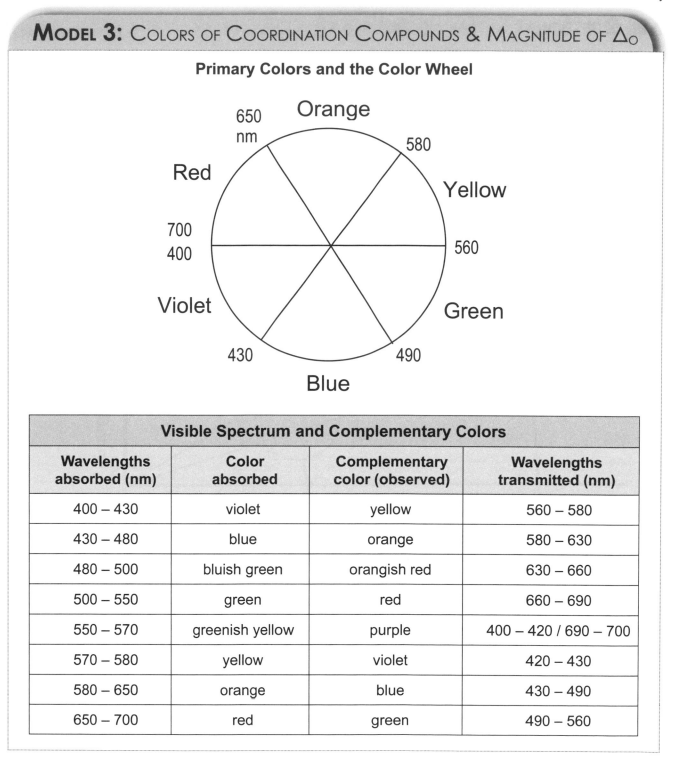

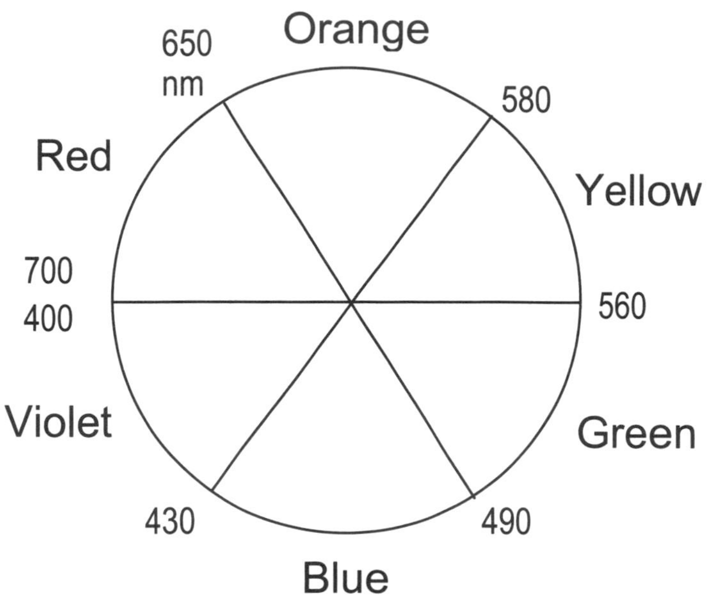

Visible Spectrum and Complementary Colors			
Wavelengths absorbed (nm)	**Color absorbed**	**Complementary color (observed)**	**Wavelengths transmitted (nm)**
400 – 430	violet	yellow	560 – 580
430 – 480	blue	orange	580 – 630
480 – 500	bluish green	orangish red	630 – 660
500 – 550	green	red	660 – 690
550 – 570	greenish yellow	purple	400 – 420 / 690 – 700
570 – 580	yellow	violet	420 – 430
580 – 650	orange	blue	430 – 490
650 – 700	red	green	490 – 560

Key Questions

9. A solution of $[CrF_6]^{3-}$ absorbs the wavelengths in the range 610 – 690 nm. What is the color of the solution?

10. A solution of $[Cr(H_2O)_6]^{3+}$ is violet. What color and wavelengths are absorbed by this complex ion?

11. A solution of $[Cr(NH_3)_6]^{3+}$ absorbs wavelengths in the range of 435 to 480 nm. What color is this solution and what color does it absorb?

12. What are examples that you have observed of magnetism and color being used for some useful purpose?

Information

In transition metals, electrons in the outer s orbitals have a lower ionization energy than d electrons. Consequently, in forming transition metal ions, the s electrons are removed first and *then* the d electrons. For example, the electron configuration of Fe^{2+} is $[Ar]3d^6$ *not* $[Ar]4s^23d^4$. Similarly, the electron configuration of Fe^{3+} is $[Ar]3d^5$.

Exercises

4. Recall that wavelength and energy are related through the equations, $E = h\nu$ and $c = \nu\lambda$, where c is the speed of light, ν is the frequency of the radiation, and λ, its wavelength. Using this information, arrange the chromium compounds in Key Questions 9, 10, and 11 in order of decreasing energy of the photons absorbed.

5. Examine the d-orbital energy level diagrams in **Model 2**, and draw similar diagrams for octahedral complex ions of each of the following:

   Fe^{3+} (weak field) Fe^{3+} (strong field)

   Ni^{2+} (weak field) Ni^{2+} (strong field)

6. Identify the number of unpaired electrons in each of the following complex ions and determine whether the ion is diamagnetic or paramagnetic:

$Ru(NH_3)_6^{2+}$ $Fe(CN)_6^{3-}$

PROBLEMS

1. The octahedral complex of SCN^- with Fe^{3+} is paramagnetic with five unpaired electrons. Does SCN^- produce a stronger or weaker ligand field than CN^-? Explain.

2. The coordination compound $Ni(H_2O)_6Cl_2$ is green (absorbs red light). The coordination compound $Ni(NH_3)_6Cl_2$ is purple (absorbs yellow-green light).

   a) Which compound absorbs light with the shorter wavelength?

   b) Which compound has the larger ligand field splitting of the d-orbitals?

   c) Which ligand produces the stronger field?

3. If you want to determine whether a given ligand is a strong-field or a weak-field ligand by experimentally determining the number of unpaired electrons from the paramagnetism, would it be better to use octahedral complexes of Cr^{2+} or Ni^{2+}? Explain.

4. The electron in the d-orbital of $Ti(H_2O)_6^{3+}$ absorbs light at 570 nm in going from a lower energy d-orbital, called t_2g, to a higher energy d-orbital, called e_g.

   a) How large is the ligand field splitting in this octahedral complex ion in units of kJ/mol?

   b) How large is the ligand field splitting compared to the first ionization energy of titanium (659 kJ/mol)?

Activity 20-1

Functional Groups

Why?

A *functional group* is a specific arrangement of atoms in an organic molecule. The functional groups determine a compound's chemical reactivity and physical properties. Functional groups are important because they make it possible to arrange a very large number of organic molecules into a much smaller number of classes. Molecules in the same class with the same functional groups undergo the same types of chemical reactions and have similar physical properties. If you are able to recognize the functional groups in a molecule, then you will be able to identify and understand the molecule's chemical reactivity and physical properties. The names of organic compounds are also directly connected to the functional groups present.

Learning Objective

- Associate the arrangement of atoms in a molecule with specific functional groups

Success Criteria

- Identify the functional groups in a molecule from its structure
- Draw a structure for a given functional group

Task

Identify the distinguishing characteristics of each functional group listed in the Model on the next page.

alkane	
alkene	
alkyne	
aromatic ring	
haloalkane	
alcohol	

ether	
amine	
aldehyde	
ketone	
carboxylic acid	
ester	
amide	

Model: Some Common Functional Groups

Name	Structure	General Formula	Example
alkane	C—H C—C	RH	CH_4, methane CH_3CH_3, ethane
alkene	C=C	$R_2C{=}CR_2$	CH_2CH_2, ethene
alkyne	C≡C	RC≡CR	CHCH, ethyne
aromatic ring		ArH	C_6H_6, benzene
haloalkane	C—X	RX	CH_3Cl, chloromethane
alcohol	C—O—H	ROH	CH_3CH_2OH, ethanol
ether	**C**—O—**C**	ROR	CH_3OCH_3, dimethylether

Model: Some Common Functional Groups (continued)

Name	Structure	General Formula	Example
amine	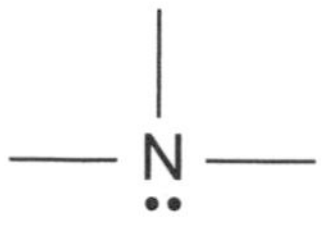	R_3N	CH_3NH_2, methylamine
aldehyde	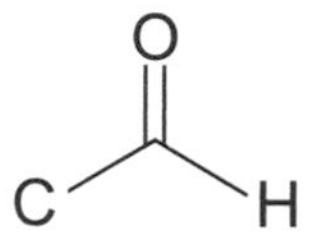	R(CO)H	$CH_3(CO)H$, acetaldehyde
ketone	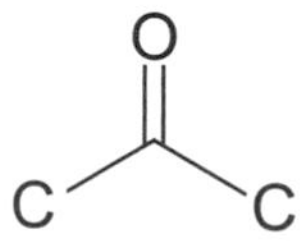	R_2CO	$(CH_3)_2CO$, acetone
carboxylic acid	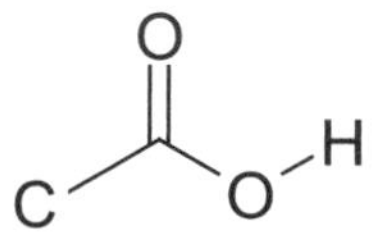	RCOOH	CH_3COOH, acetic acid
ester	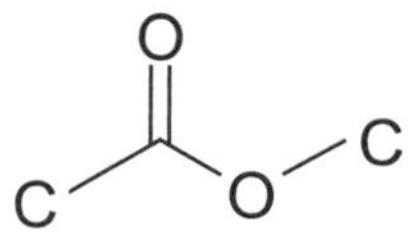	R(CO)OR	$CH_3(CO)OCH_3$, methyl acetate
amide	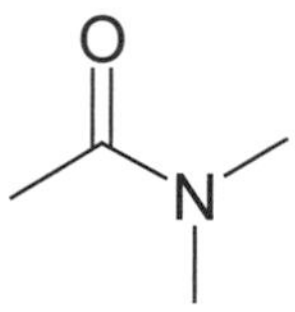	$R(CO)NR_2$	$CH_3(CO)NH_2$, acetamide

Information

Line structures are used to represent the molecules. In a line drawing, each line represents a single bond. A carbon atom is implicit where two lines come together or a line ends. Each carbon atom needs four bonds, so any missing bonds are actually implicit bonds to hydrogen atoms.

Exercises

1. Identify the functional groups in the following compounds.

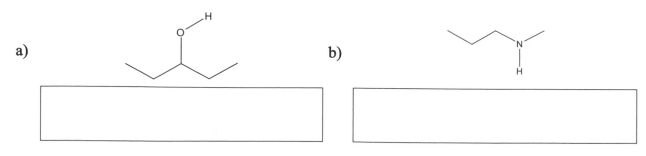

a) 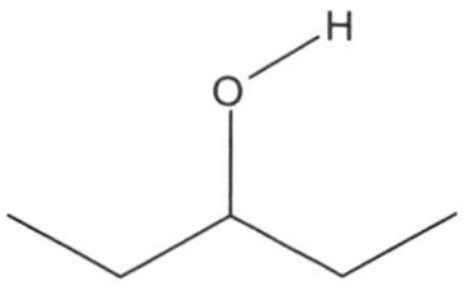

b) 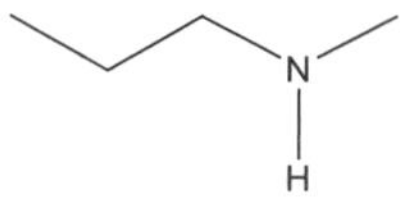

c) 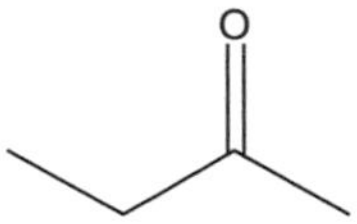

d) 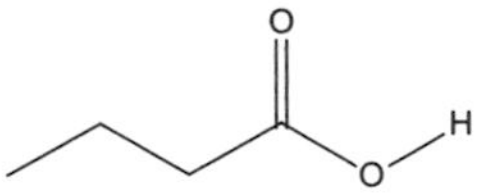

2. Draw the line structure of molecules containing the following functional groups.

a) alkyne

b) ether

c) aldehyde

d) ester

e) amide

3. Write the molecular formula and identify all the structural groups in each of the following compounds.

   a) Lactic acid, found in milk products

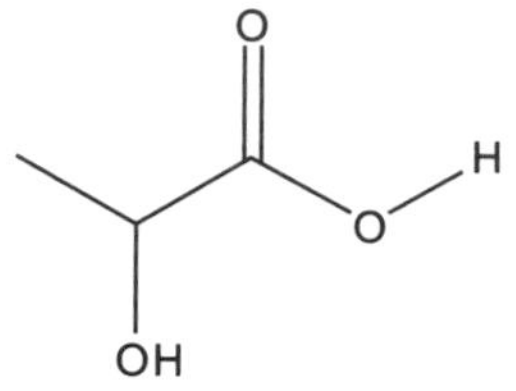

   b) Ephedrine, a decongestant

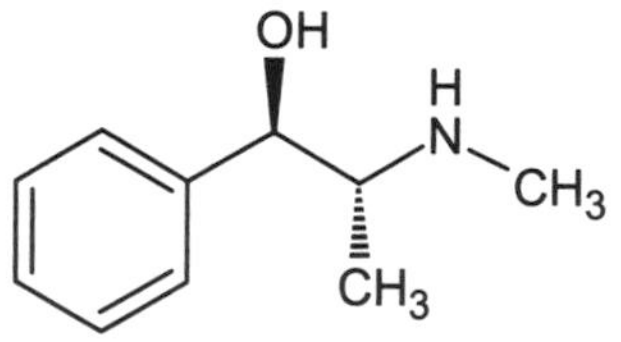

   c) Ethyl acetate, nail polish remover

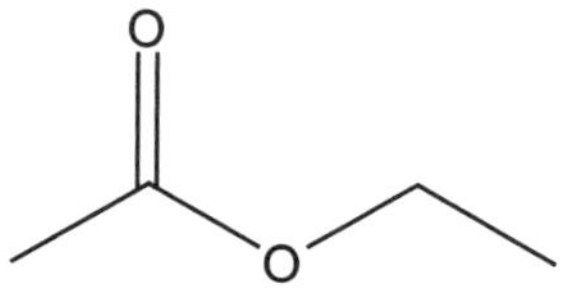

   d) Aspartame, an artificial sweetener

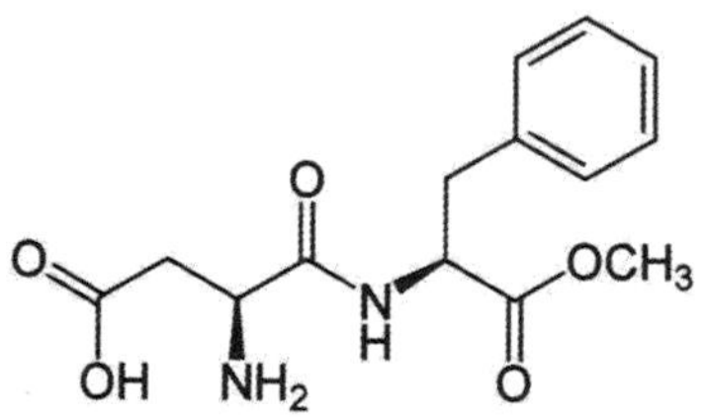

   e) Sucrose, table sugar

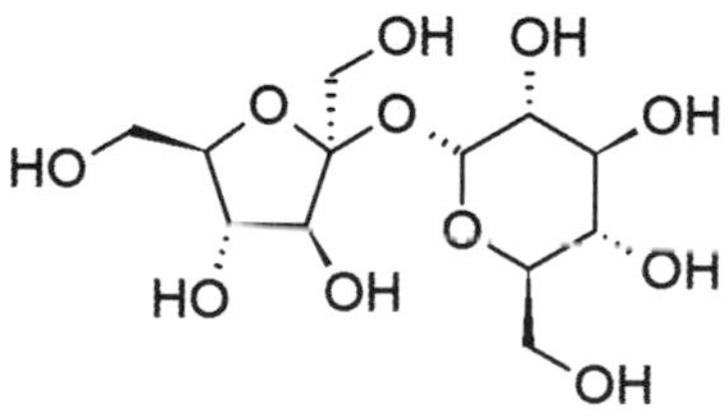

   f) Nylon, a synthetic polymer

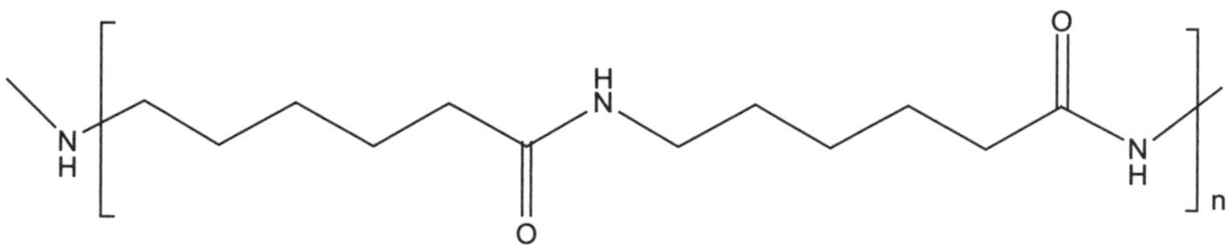

ACTIVITY 20-2

Polymer Chemistry

WHY?

Polymers are large molecules that are built from repeating units called *monomers*. Because the properties of polymers are extremely diverse, they have widespread and increasing applications. They are key components in living organisms and have many applications in daily life, medicine, technology, and industrial processes. You should know something about the properties of polymers and how polymers are made because you encounter them daily and the jobs of many chemists involve working with polymers.

LEARNING OBJECTIVES

- Identify reactions that are used to synthesize polymers
- Summarize information on polymer structure, properties, and uses

SUCCESS CRITERIA

- Ability to identify polymerization reactions and properties of polymers
- Ability to write the reactants and products of polymerization reactions

PREREQUISITES

- **Activity 03-1:** *Molecular Representations*
- **Activity 04-3:** *Introduction to Acid – Base Reactions*

INFORMATION

Addition Polymers

Small molecules with multiple bonds can come together as *monomers* to form larger molecules called *polymers*. During the polymerization process, the multiple bonds are lost and new, single bonds between the monomers are formed. Each monomer is simply added to the end of a growing chain, so these molecules are called *addition polymers*. The formation of polyethylene from ethylene is an example of addition polymerization.

C_2H_4 ethylene $(C_2H_4)_n$ polyethylene

Contributed by Joseph Lauher, Stony Brook University

Key Questions

1. What is the relationship between a monomer and a polymer?

2. What are the characteristics of addition polymerization?

Exercises

1. The following molecules also undergo addition polymerization via the ethylene double bond. Draw the respective polymers.

a) 

acrylonitrile

b) 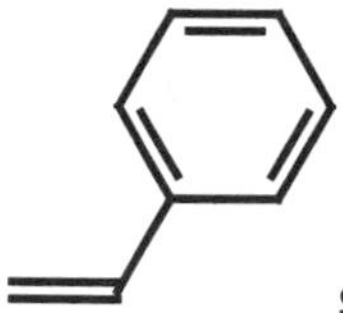

styrene

c) 

propene (or propylene)

2. What monomers would you use to form the following addition polymers?

a) 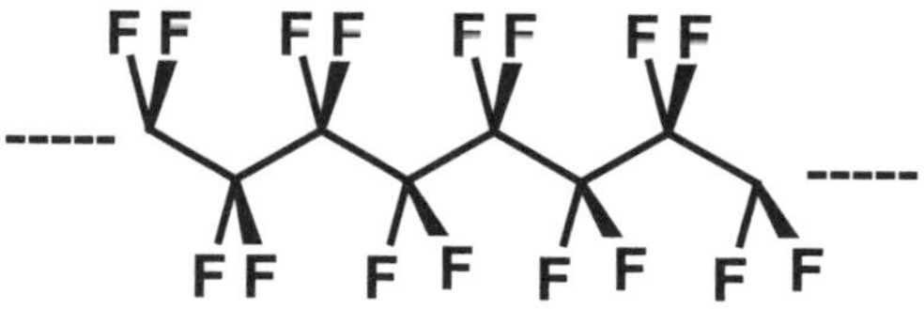

polytetrafluoroethylene (PTFE) "Teflon"

b) 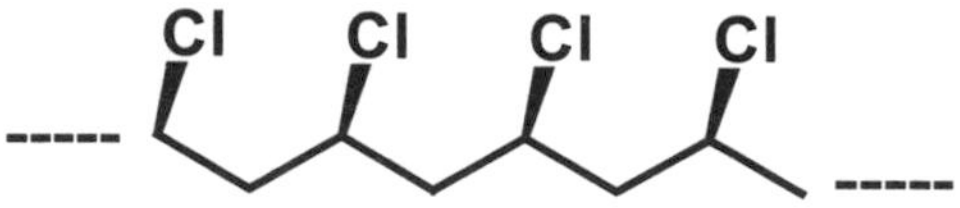

polyvinylchloride (PVC)

PROBLEMS

1. When the molecule styrene is polymerized, sometimes a small amount of a compound called divinylbenzene is added. This addition makes the resulting polymer harder and stronger via a process called *crosslinking*. Explain crosslinking in terms of addition reactions involving these two reactants. Illustrate your explanation with a drawing of the cross-linked polymer.

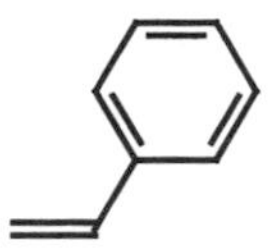

styrene

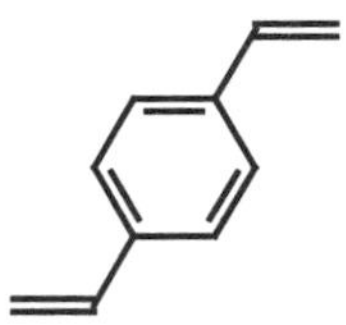

divinylbenzene

2. a) Natural rubber is a polymer of the molecule called isoprene. Polyisoprene is a linear chain with a methyl side group next to a double bond in the chain. Draw the polymer in box 1, below.

   b) Charles Goodyear discovered a process called *vulcanization* that led to an increase in the strength and durability of rubber by crosslinking the polymer with the element sulfur. The double bonds in the polyisoprene chain are not affected by the cross linking. Draw the crosslinked polymer in box 2, below. (If you can't figure this out, maybe you can find pictures in your text or another resource.)

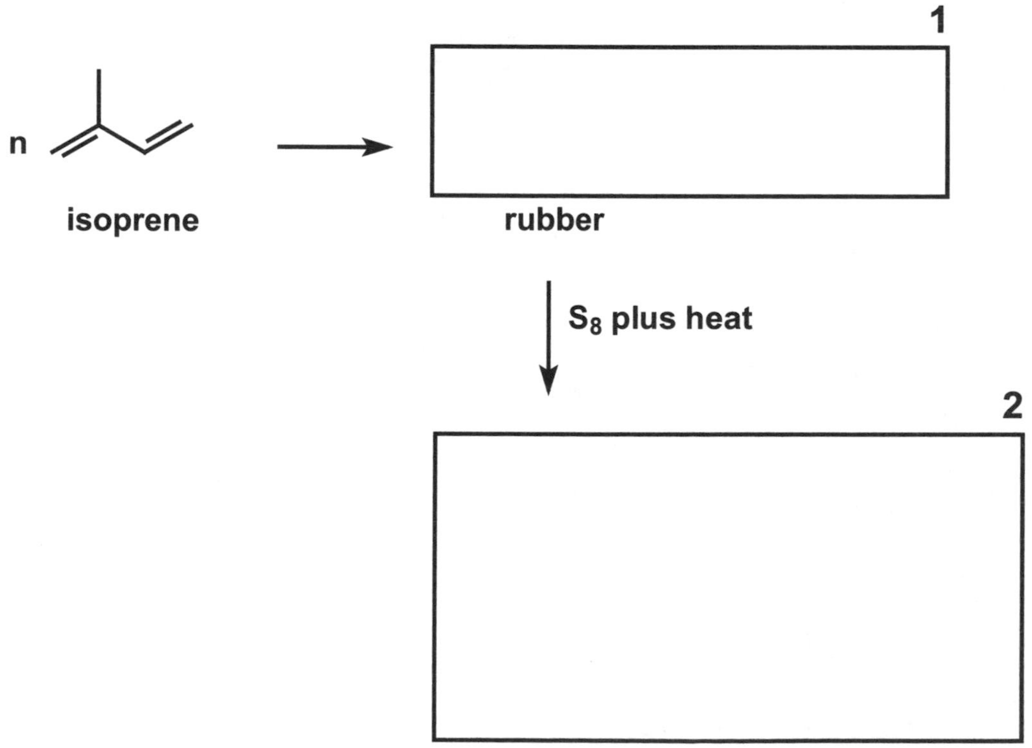

Information

Condensation polymers

In a condensation polymer, monomers form bonds between their functional groups by eliminating a small molecule such as water or HCl. Two common examples given below are amides and esters.

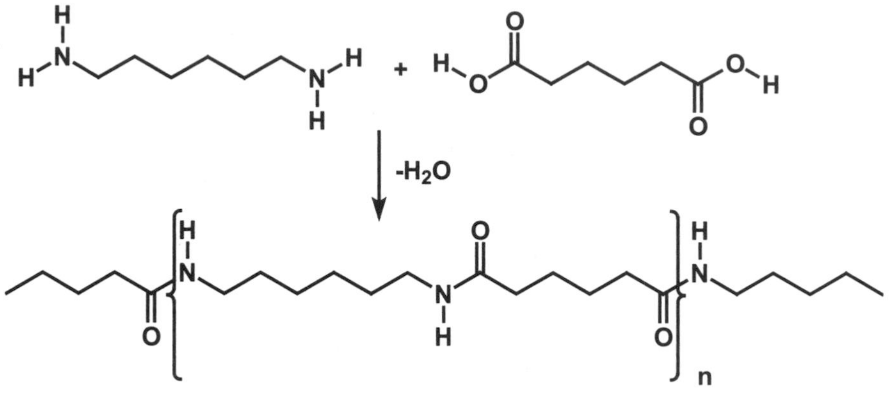

nylon 66 - a polyamide

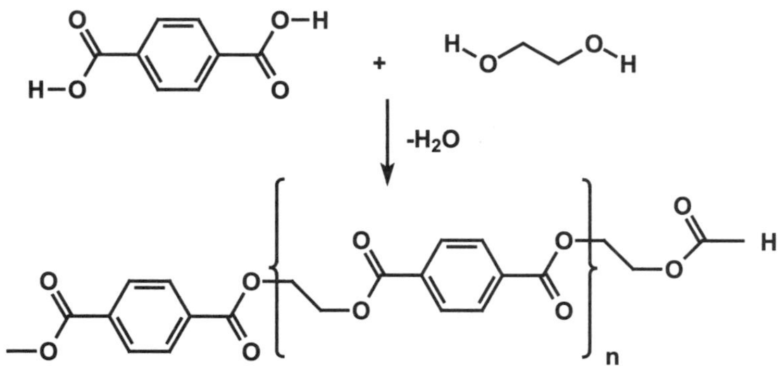

a polyester

Exercises

3. Draw the polymers that would result from the following condensation reactions, and identify the small molecule that is eliminated in the reaction.

a) 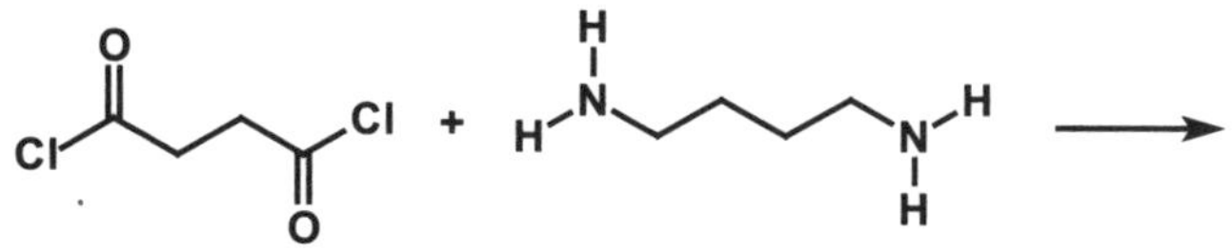

b) 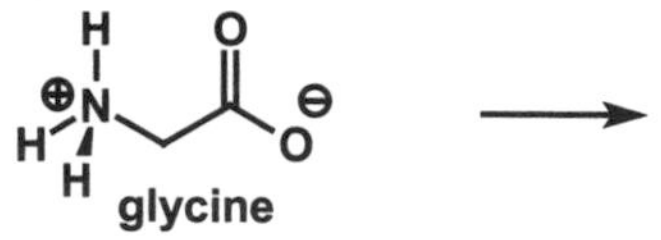

c) 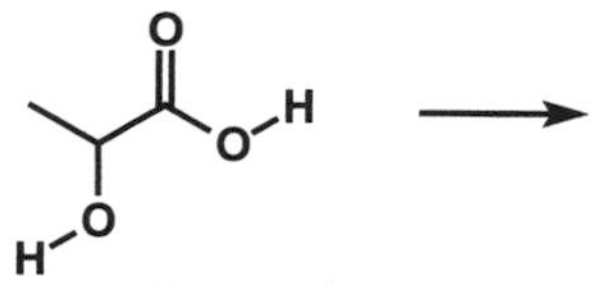

lactic acid

4. What monomers would give the following condensation polymers?

a) 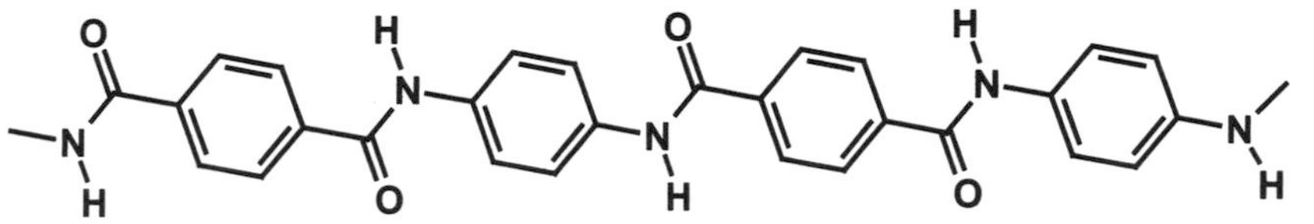

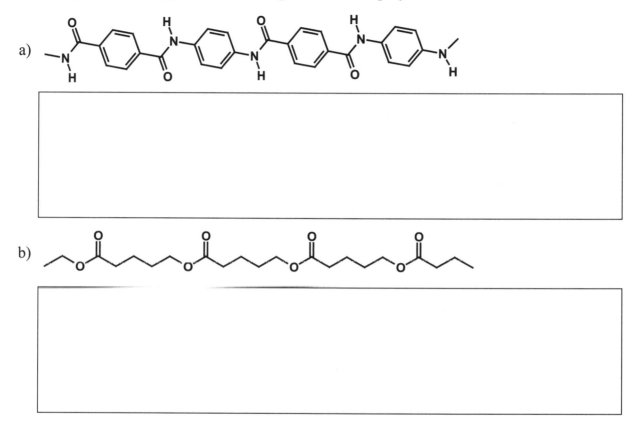

b) 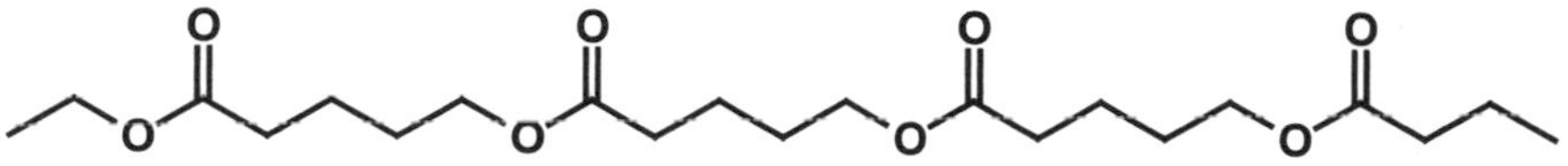

Key Questions

3. What structural features must a molecule have in order to form a polymer by a condensation reaction?

4. What structural features must a molecule have in order to form a polymer by an addition reaction?

5. What features in the structure of a polymer would make the material very strong and rigid?

6. Why is it that, when heated, polymers do not melt and become liquid, in the same way that ice melts and becomes water?

ACTIVITY 21-1

Amino Acids and Proteins

WHY?

Proteins constitute about 50% of the dry weight of most cells, and are the most structurally complex macromolecules known. Each type of protein has its own unique structure and function. Knowledge of the structures of these molecules is essential for continuing studies in chemistry and biology.

LEARNING OBJECTIVES

- Understand how amino acids combine to form proteins
- Understand the different properties of amino acids

SUCCESS CRITERIA

- Identify the two major types of protein secondary structures: α-helices and β-sheets
- Ability to draw structures of amino acids and peptides

PREREQUISITES

- **Activity 03-1:** *Molecular Representations*
- **Activity 20-2:** *Polymer Chemistry*

INFORMATION

Polymers are any kind of large molecule made of repeating identical or similar subunits called *monomers*. Amino acids are the monomers that are used to build proteins. To form proteins, the amino acids are linked by a condensation reaction that produces an amide linkage (-CO-NH-) and eliminates water. The amide linkage in proteins is also called a *peptide linkage*, and the chain of amino acids is also known as a *polypeptide*, or simply a *peptide*. Generally peptides consist of no more than 50 to 100 amino acids, while most proteins typically consist of 100 to 1000 amino acids. Some proteins contain only one polypeptide chain while others, such as hemoglobin, contain several polypeptide chains all twisted together. The sequence of amino acids in each polypeptide or protein is unique and consequently each protein has its own unique 3-D shape or *native conformation*.

If even one amino acid in the sequence is changed, the protein's ability to function can be impared. For example, sickle cell anemia is caused by a change in only one nucleotide in the DNA sequence that causes just one amino acid in one of the hemoglobin polypeptide molecules to be different. Because of this, the whole red blood cell ends up being deformed and is unable to carry oxygen properly.

The amino acids are known as *α-amino acids* because (with the exception of proline) they all have the same structure: a primary amino group and a carboxylic acid group attached to the same α-carbon that carries a side chain.

Contributed by Frank Fowler and Andisheh Abedini, Stony Brook University

All amino acids have the same general structural formula:

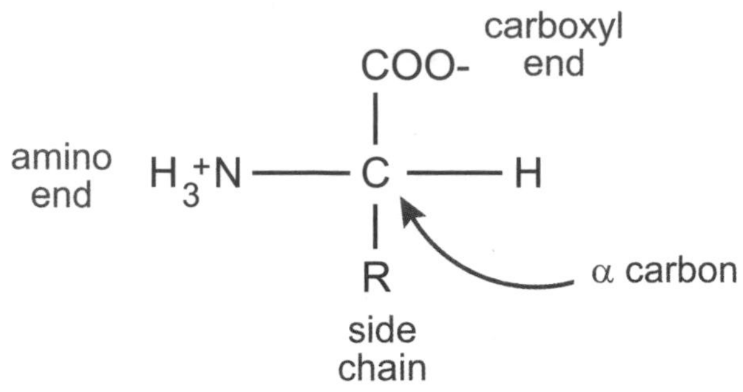

A protein's primary structure (1°) is the amino acid sequence of its polypeptide chain(s).

A protein's secondary structure (2°) is the local spatial arrangement of a polypeptide's backbone atoms without regard to the conformations of its side chains.

A protein's tertiary structure (3°) refers to the three-dimensional structure of an entire polypeptide. The tertiary structure is stabilized by the interactions among the individual side groups (R) on the amino acids comprising the protein. Many proteins are composed of two or more polypeptide chains, referred to as *subunits*, which associate through noncovalent interactions and, in some cases, disulfide bonds. A protein's quaternary structure (4°) refers to the spatial arrangement of these subunits.

Key Questions

1. What monomers are the building blocks of proteins?

2. Which type of polymerization reaction is used to attach the monomers to each other to produce peptides and proteins?

3. What is the name of the linkage or bond formed between the monomers in a protein?

4. What is it about their structure that gives different α-amino acids different properties?

5. What is it about their structure that gives different proteins different properties?

6. If there are 20 different amino acids, how many different peptides can be produced with a chain length of 50 amino acids?

7. Why is it important to recognize that proteins have a primary, secondary, tertiary, and quaternary structures?

Exercises

1. The various amino acids are usually classified according to the polarities of their side chains, **R**, which are attached to the α-carbon. Use your text or other resource as a reference in completing the following exercises.

   a) Write the names and three-letter abbreviations for the amino acids with nonpolar side chains.

   b) Write the names and three letter abbreviations for the amino acids with uncharged polar side chains.

   c) Write the names and three letter abbreviations for the amino acids with charged polar side chains.

2. In the physiological pH range (around 7), both the carboxylic acid and the amino groups of the amino acids are completely ionized, meaning that they carry charges. These ions have both a positive and a negative charge and are called *zwitterions*, which is from the German meaning hybrid ions. Draw the structure of an amino acid in the zwitterion form. Include the carboxylic acid group and the amino group, designate the side chain on the α-carbon as R, and label the location of the + and – charges.

3. Write the reaction equation using structural formulas to show how alanine and glycine react to form the dipeptide alanylglycine.

4. Draw the structural formula of three peptide-bonded amino acids using R1, R2, and R3 to designate the side chains.

5. Draw the following peptides in their predominant ionic forms at pH 7:

   a) Phe-Met-Arg

b) Gln-Ile-His-Thr-Arg

c) Gly-Pro-Tyr-Cys-Lys

PROBLEMS

1. One major type of protein secondary structure is the α-helix. α-helices are held together by hydrogen bonds and have a right-handed turn every 3.6 residues. The hydrogen bonds of α-helices are arranged such that the peptide C=O bond of the nth residue points along the helix towards the peptide N-H group of the $(n+4)^{th}$ residue. The R groups of the α-helices all project outward and downward from the helix to avoid steric interference with the polypeptide backbone and with each other.

   Draw an example of an α-helix. On the helix, show the number of amino acids per turn as squares; designate where hydrogen bonds would be between C=O and –NH. Also show where the side chains would be. (***Hint:*** Find a figure in your text or other resource.)

2. β-sheets are the second major form of secondary structure. In a β-sheet, hydrogen bonding occurs between neighboring polypeptide chains rather than within one as in α-helices.

   a) Use your text or other resource to help you identify the two major types of β-sheets and explain the difference between the two.

b) Draw antiparallel and parallel β-sheets using arrows to point toward the direction of chains.

3. In your own words, define what protein tertiary and quaternary structure entails.

ACTIVITY 22-1

Nuclear Chemistry: Binding Energy

WHY?

A nuclide is an atomic nucleus with a specific number of protons and neutrons. The binding energy of a nuclide is the energy required to separate the nuclide into its component protons and neutrons. Binding energies of nuclides determine the energy released in nuclear reactions and provide insights regarding the relative stability of elements and their isotopes.

WHAT DO YOU THINK?

Centuries ago, a major goal of alchemists was to turn lead into gold. Do you think this is possible? Explain why or why not.

LEARNING OBJECTIVE

- Understand the origin of energy produced in nuclear reactions

SUCCESS CRITERIA

- Explain the source of energy in a nuclear reaction
- Correctly calculate nuclear binding energies and the energy released in nuclear reactions

PREREQUISITES

- **Activity 02-1:** *Atoms, Isotopes, and Ions*
- **Activity 14-2:** *Entropy of the Universe and Gibbs Free Energy*

PROCEDURE

Calculation of the Binding Energy per Nucleon

Since the advent of Albert Einstein's Theory of Special Relativity, mass and energy are considered to be different manifestations of the same quantity. The equation $E = mc^2$ (where c is the speed of light) provides the conversion between mass, m, and energy, E. In most situations, e.g., chemical reactions, energy changes are so small that corresponding changes in mass cannot be observed. The binding energies of nuclides are very much larger, so it is useful to view the binding energies as changes in mass. Consequently the binding energy of a nuclide, E_b, and the binding energy per nucleon, $E_{b/A}$, can be calculated by the following procedure, where A is the mass number. Step (5) in this procedure is very important because to see how strongly one nucleon is held in the nuclide, it is necessary to divide the total binding energy of the nuclide by the number of nucleons comprising it.

(1) $E_b = E_{separated\ nucleons} - E_{nuclide}$

(2) $E_b = m_{separated\ nucleons}\ c^2 - m_{nuclide}\ c^2$

(3) $E_b = (m_{separated\ nucleons} - m_{nuclide})\ c^2$

(4) $E_b = \Delta mc^2$

(5) $E_{b/A} = \frac{E_b}{A}$

Analysis

How to calculate binding energies

1. Different books disagree over which isotope of iron is the most stable.
   a) Examine the Procedure above, and describe in words what you need to do to calculate the binding energy per nucleon for an isotope of iron.

   b) Identify the concepts or ideas that you need to use in this calculation and explain why you need to use them.

2. Calculate the binding energy per nucleon for iron-56 and iron-58. Place these points, along with the values you calculated, on the graph in the Model, and identify the isotope that is more stable. For this calculation you need precise values for the mass of a proton, a neutron, and the iron isotopes. Remember that 1 J = 1 kg m^2/s^2, and note that the units on the graph are 10^8 kJ/mol.

proton mass = 1.007276 g/mol, neutron mass = 1.008665 g/mol,

Fe-56 mass = 55.934994 g/mol, Fe-58 mass = 57.933275 g/mol

MODEL: BINDING ENERGIES OF NUCLEONS

Figure 1

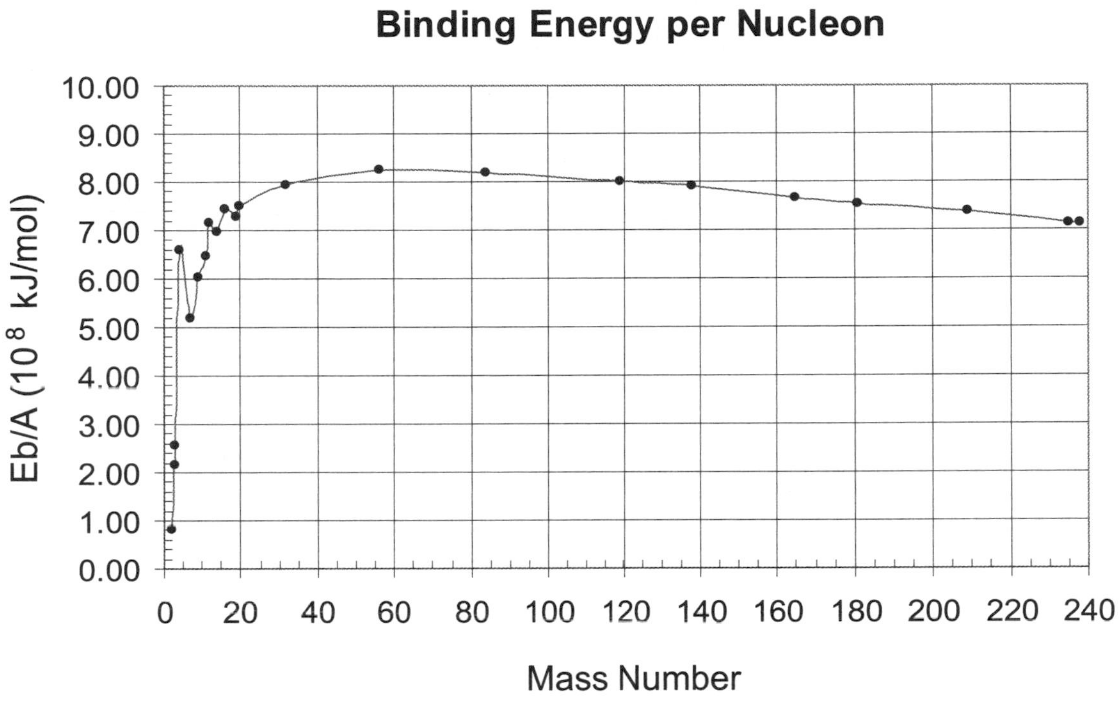

KEY QUESTIONS

1. Of all isotopes in the Model, which has the smallest binding energy per nucleon?

2. Why do you think the binding energy per nucleon is small when the mass number is very small?

3. Using the ideas of a repulsive electromagnetic force and an attractive nuclear force, why do you think the binding energy per nucleon first increases and then decreases as the number of nucleons increases past some intermediate value?

4. Which nuclides are more stable: those with a large binding energy per nucleon or those with a small binding energy per nucleon? Explain.

5. Is the reaction that converts a less stable nuclide to a more stable nuclide exothermic or endothermic? Would you expect the free energy change for this reaction to be positive or negative? Explain.

6. In view of your answer to Key Question 5, which of those isotopes in the Model would you expect to be most abundant at equilibrium? Explain.

7. Why is the distribution of isotopes not the equilibrium distribution, e.g., why do we still have hydrogen, helium, and uranium?

8. Will a nuclide with a small mass number become more stable by combining with another nuclide or by splitting into smaller nuclides? Explain.

9. Will a nuclide with a large mass number become more stable by combining with another nuclide or by splitting into smaller nuclides? Explain.

10. There are two kinds of atomic bombs: hydrogen and uranium. A hydrogen bomb is much more destructive. Why do you think more energy is released from a hydrogen bomb than a uranium bomb?

Information

All elements are thermodynamically unstable with respect to iron because the free energy change required to produce iron from these elements is negative. All elements should spontaneously convert to iron but do not do so when the activation energy for such conversion is large. Light nuclides can come together to form more stable heavier nuclides. This process is called *fusion*, which is the source of energy in the sun. Heavy nuclides can split to form more stable lighter nuclides. This process is called *fission*, which is the source of energy in nuclear power generation stations.

Exercises

1. Calculate the energy released when one mole of ^2H is converted into ^3He. The nuclear reaction equation is $2\ ^2H \longrightarrow\ ^3He$ + neutron. The nuclide masses are ^2H = 2.0141 g/mol, ^3He = 3.0160 g/mol.

2. How much energy is released per mole of uranium in the following nuclear reaction? The nuclide masses in g/mol are U (235.0439), Mo (99.9076), and Sn (133.9125).

$$^{235}U + \mathrm{n} \longrightarrow {}^{100}Mo + {}^{134}Sn + 2n$$

What do you think now?

Is it possible to turn lead into gold? If no, explain why not. If yes, explain how it might be done. In your explanation, mention whether gold is more or less stable than lead.

Activity 22-2

Radioactivity

Why?

Not all elements and isotopes are stable. Radioactivity is the spontaneous emission of particles or energy from the nucleus of an unstable isotope. Such isotopes are called *radioisotopes*. These spontaneous nuclear reactions produce more stable elements and isotopes from those that are less stable. Nuclear chemistry is the subdiscipline of chemistry that studies these reactions and utilizes them in research, technology, and medicine. An understanding of nuclear chemistry is essential in discussing issues associated with radioactivity, nuclear power generation, disposal of radioactive waste, and the intelligent use of radiation in medicine.

What Do You Think?

Are nuclear reactions similar to chemical reactions? Identify any similarities and differences.

Learning Objective

- Understand what happens in the radioactive decay of an isotope

Success Criterion

- Write correct nuclear reaction equations to describe how nuclei change by fission, fusion, and the emission or capture of particles

Prerequisites

- **Activity 04-1:** *Balanced Chemical Reaction Equations*
- **Activity 22-1:** *Nuclear Chemistry: Binding Energy*

Tasks

1. Examine the Model on the following page, and for each example, identify the quantities that are conserved in the nuclear reaction. The *Key Questions* will help you with this identification.

2. Using the conservation rules that you have identified, complete the second example that is given in each row of the Model.

Information

Just as was done for an atomic symbol, the symbol for a nuclide includes the symbol for the element, a superscript for the mass number, A, which is equal to the total number of protons and neutrons, and a subscript for the atomic number, Z, which specifies the number of protons in the nuclide and also the charge of the nuclide. Since superscripts and subscripts are also used in symbols for subatomic particles like electrons and positrons, the subscript Z now is called the *charge number*.

Model: Spontaneous Nuclear Reactions

Process	Symbol	Δ Z	Δ A	Examples
alpha particle emission	α or ${}^{4}_{2}He$	–2	–4	${}^{238}_{92}U \rightarrow {}^{234}_{90}Th + \alpha$ ☐ $\rightarrow {}^{222}_{86}Rn + \alpha$
beta particle emission	β or ${}^{0}_{-1}e$	+1	0	${}^{131}_{53}I \rightarrow {}^{131}_{54}Xe + \beta$ ☐ $\rightarrow {}^{32}_{16}S + \beta$
positron emission	β⁺ or ${}^{0}_{+1}e$	–1	0	${}^{22}_{11}Na \rightarrow {}^{22}_{10}Ne + \beta^{+}$ ☐ $\rightarrow {}^{15}_{7}N + \beta^{+}$
neutron emission	${}^{1}_{0}n$	0	–1	${}^{12}_{4}Be \rightarrow {}^{11}_{4}Be + {}^{1}_{0}n$ ☐ $\rightarrow {}^{4}_{2}He + {}^{1}_{0}n$
gamma ray emission, excited state decays	γ	0	0	${}^{57}_{26}Fe^{*} \rightarrow {}^{57}_{26}Fe + \gamma$ ☐ $\rightarrow {}^{119}_{50}Sn + \gamma$
electron capture (K-capture)	${}^{0}_{-1}e$	–1	0	${}^{7}_{4}Be + {}^{0}_{-1}e \rightarrow {}^{7}_{3}Li$ ☐ $+ {}^{0}_{-1}e \rightarrow {}^{40}_{18}Ar$
fission		various	various	${}^{236}_{92}U \rightarrow {}^{141}_{56}Ba + {}^{92}_{36}Kr + 3\,{}^{1}_{0}n$ ☐ $\rightarrow {}^{103}_{42}Mo + {}^{131}_{50}Sn + 2\,{}^{1}_{0}n$
fusion		various	various	${}^{2}_{1}H + {}^{2}_{1}H \rightarrow {}^{3}_{2}He + {}^{1}_{0}n$ ☐ $+ {}^{3}_{2}He \rightarrow {}^{4}_{2}He + {}^{0}_{+1}e$

Key Questions

1. What information is provided by the superscripts and subscripts on the symbols used in the Model?

2. What is the change in the charge number and what is the change in the mass number when an alpha particle is emitted from a nuclide?

3. What is the change in the charge number and what is the change in the mass number when a beta particle is emitted from a nuclide?

4. What is the relationship between the mass numbers of the reactants and the mass numbers of the products in a nuclear reaction?

5. What is the relationship between the charge numbers of the reactants and the charge numbers of the products in a nuclear reaction?

6. Can a nuclear reaction appear to turn a proton into a neutron or a neutron into a proton? If it cannot, explain why not. If it can, describe how.

7. Using your knowledge about chemical reactions, draw an analogy to identify what in a nuclear reaction determines whether it is spontaneous or not.

8. Using your knowledge about chemical reactions, draw an analogy to identify what determines whether a nuclear reaction has a small or large rate constant.

Exercises

1. Write an equation describing the radioactive decay of each of the following isotopes. The type of decay is given in parentheses.

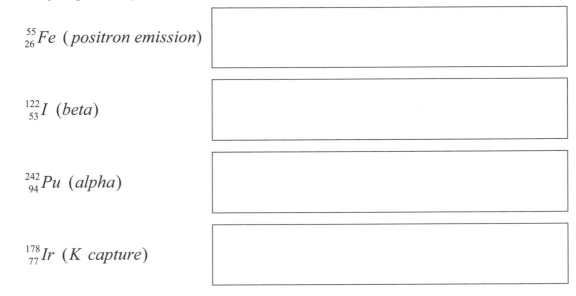

$^{55}_{26}Fe$ (*positron emission*)

$^{122}_{53}I$ (*beta*)

$^{242}_{94}Pu$ (*alpha*)

$^{178}_{77}Ir$ (*K capture*)

What Do You Think Now?

Identify the similarities and differences between chemical reactions and nuclear reactions.

Activity 22-3

Rates of Radioactive Decay

What Do You Think?

Radioactive iodine-131 is used to treat thyroid cancer. A patient is given 20.0 mg of iodine-131 in the form of NaI, and after 8 days only 10 mg remain. How long, from the time of the initial dose, do you think it will it take until only 5 mg remain?

a) 12 days

b) 16 days

c) 20 days

d) cannot be determined from the information given

Learning Objectives

- Understand how the level of radioactivity decreases with time
- Understand how radioactivity can be used to determine the age of materials

Success Criteria

- Relate the amount of radioactive material remaining after some period of time to the half-life of the radioactive isotope and the rate constant for the decay
- Estimate the age of materials from their radioactivity

Prerequisites

- **Activity 13-1:** *Rates of Chemical Reactions*
- **Activity 22-2:** *Radioactivity*

Tasks

1. Examine the Model and complete Table 1 by entering N, the number of unstable nuclides remaining at each point in time, and ln(N), in the last two columns, respectively.

MODEL: TIME EVOLUTION OF NUCLEAR DECAY

In the chart below, ten unstable nuclides are represented by white circles. They decay spontaneously by some mechanism to produce stable nuclides, which are indicated by dark circles.

Table 1

Time (mins)	Original and Product Nuclides	N	ln(N)
0	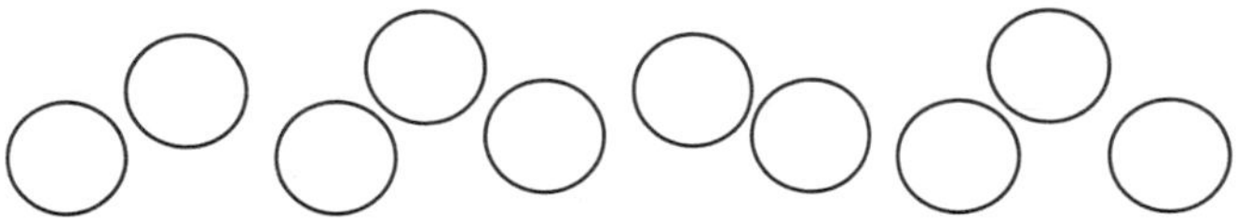	10	2.30
2.6	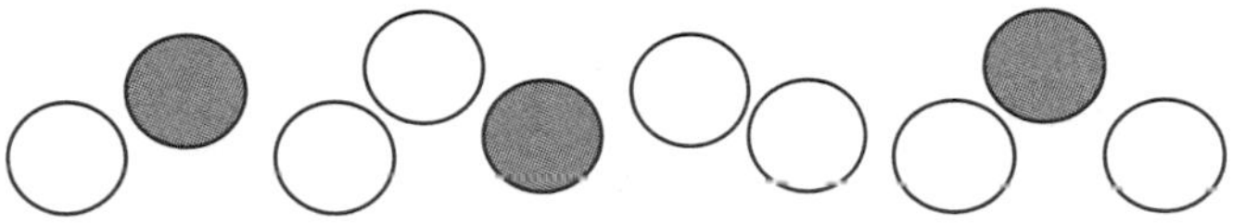		
5.0	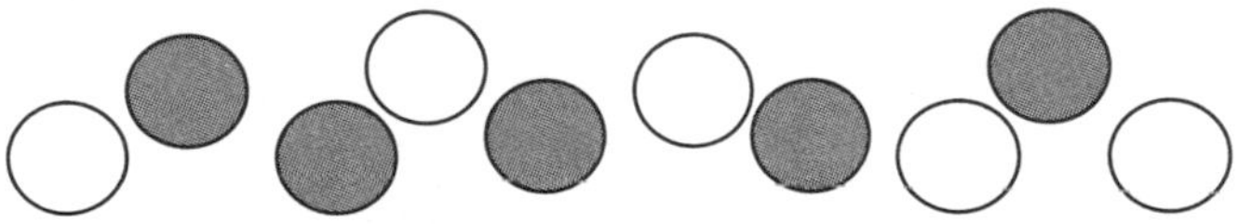		
8.7	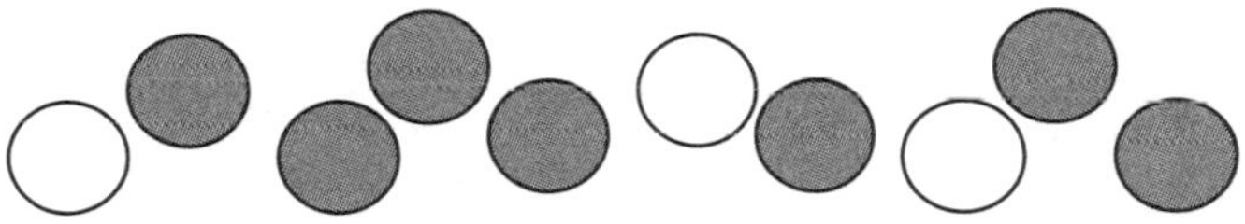		
11.6	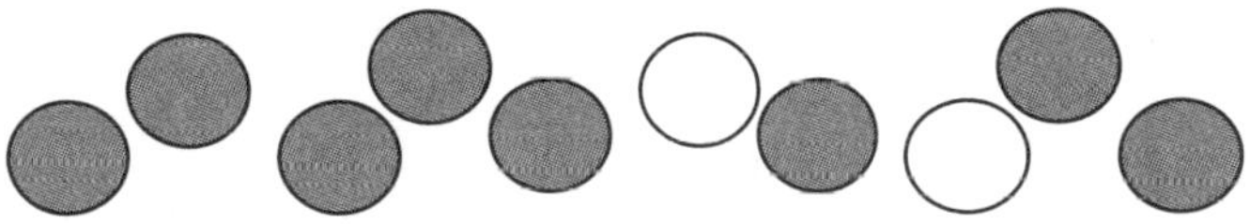		
16.6	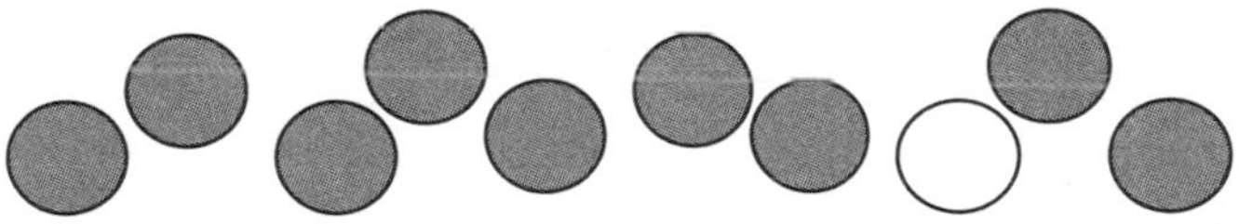		

2. Make two graphs of the data (see the next page): Graph 1 with N plotted on the y-axis and time plotted on the x-axis, and Graph 2 with ln(N) plotted on the y-axis and time plotted on the x-axis. (Remember that high quality graphs have titles, labels on the axes, data points shown with small circles around them, and a smooth line drawn through the data points.)

3. On your graphs, mark the points on the x-axis where the fraction remaining is equal to $\frac{1}{2}$, $\frac{1}{4}$, and $\frac{1}{8}$ of the initial amount.

Graph 1

Graph 2

KEY QUESTIONS

1. From Graph 1, how long does it take for half of the radioactive nuclides in the Model to decay? This time is called the *half-life*.

2. Does the time it takes for half of the radioactive nuclides to decay depend on the point in time taken as the starting point? Explain how the information in Graph 1 supports your answer.

3. How can the data in the Model be used to identify radioactive decay as one of the following types of reactions: zero order, first order, or second order?

4. What is the mathematical equation that represents the straight line that can be drawn through the data points in Graph 2?

5. How can the rate constant for this radioactive decay be obtained from Graph 2?

Exercises

Use the following information for Exercises 1–4: Two radioactive isotopes, A and B, have decay rate constants k_A and k_B, respectively, where k_A is larger than k_B. N_{A_0} and N_{B_0} are the number of nuclides present at t = 0 for each of these isotopes. The decay rate R is the number of decay events per second, where

$$R = \Delta N/\Delta t = -k\,N$$

1. Sketch one graph with a curve for $N_A(t)$ and a curve for $N_B(t)$ to compare the number of radioactive nuclides present as a function of time for both of these isotopes. Start at t = 0 with the same number of A isotopes and B isotopes. (It might be helpful to look at the graphs you constructed, based on the information in the Model, and think about how they would be different if the decay rate were faster.)

2. Sketch one graph with a curve for $\ln[N_A(t)]$ and a curve for $\ln[N_B(t)]$ as a function of time to compare these functions for the two isotopes. Again, start at t = 0 with same number of A isotopes and B isotopes.

3. Identify the isotope A or B with the longer half-life.

4. Identify the isotope A or B with the faster decay rate when the amounts of A and B present are equal.

5. Using your answer to Key Question 4, derive the following relationship between the half-life and the rate constant: $t_{1/2} = \ln(2)/k$.

6. Derive your answer to Key Question 4 from the fact that the data in Graph 1 is described by an exponential function: $N = N_o e^{-kt}$

7. Iodine-123 has a half-life of 13.3 hours. Using your answer to Exercise 5, calculate a value for the decay rate constant, k, in units of s^{-1}.

8. Determine the half-life of a radioactive nuclide if, after 2.5 hrs, only 1/32 of the initial amount remains unchanged.

9. Estimate how many half-lives must pass before less than 1% of a radioisotope remains.

What do you think now?

1. A patient is given 20.0 mg of iodine-131 in the form of NaI. After 8 days only 10 mg remain. How long, from the time of the initial dose, do you think it will it take until only 5 mg remain?

   a) 12 days

   b) 16 days

   c) 20 days

   d) cannot be determined from the information given

2. How long will it take until only 0.1% of the iodine-131 remains?

PROBLEMS

1. A 0.133 pg sample of iron-59 produces 242 β-particles per second. What is the half-life of this isotope?

2. A wooden tool is found at an archaeological site. Estimate the age of the tool using the following information: A 100 gram sample of the wood emits 1120 β-particles per minute from the decay of carbon-14. The decay rate of carbon-14 in living trees is 15.3 per minute per gram. Carbon-14 has a half-life of 5730 years.

3. Potassium-40 decays into argon-40 by electron capture with a half-life of 1.28×10^9 years. This decay forms the basis of the potassium-argon dating method that is used with igneous rocks. This method assumes that no argon was trapped in the rock when it was formed from the molten material.

   A meteorite was found to have equal amounts of ^{40}K and ^{40}Ar.

   a) Of the potassium-40 that was originally present, what fraction has decayed?

b) How long ago was this meteorite formed?

c) Another meteorite had a ^{40}K to ^{40}Ar ratio of 0.33. How long ago was this meteorite formed?

Appendix A

Units and Physical Constants

International System of Units (SI)

http://physics.nist.gov/cuu/Units

SI Base Units

Base quantity	Name	Symbol
length	meter	m
mass	kilogram	kg
time	second	s
electric current	ampere	A
thermodynamic temperature	kelvin	K
amount of substance	mole	mol
luminous intensity	candela	cd

SI Prefixes

Factor	Name	Symbol	Factor	Name	Symbol
10^{24}	yotta	Y	10^{-1}	deci	d
10^{21}	zetta	Z	10^{-2}	centi	c
10^{18}	exa	E	10^{-3}	milli	m
10^{15}	peta	P	10^{-6}	micro	μ
10^{12}	tera	T	10^{-9}	nano	n
10^{9}	giga	G	10^{-12}	pico	p
10^{6}	mega	M	10^{-15}	femto	f
10^{3}	kilo	k	10^{-18}	atto	a
10^{2}	hecto	h	10^{-21}	zepto	z
10^{1}	deka	da	10^{-24}	yocto	y

Definitions of SI Base Units

Unit of length	**meter**	The meter is the length of the path travelled by light in vacuum during a time interval of 1/299 792 458 of a second.
Unit of mass	**kilogram**	The kilogram is the unit of mass; it is equal to the mass of the international prototype of the kilogram.
Unit of time	**second**	The second is the duration of 9 192 631 770 periods of the radiation corresponding to the transition between the two hyperfine levels of the ground state of the cesium 133 atom.
Unit of electric current	**ampere**	The ampere is that constant current which, if maintained in two straight parallel conductors of infinite length, of negligible circular cross-section, and placed 1 meter apart in vacuum, would produce between these conductors a force equal to 2×10^{-7} newton per meter of length.
Unit of thermodynamic temperature	**kelvin**	The kelvin, unit of thermodynamic temperature, is the fraction 1/273.16 of the thermodynamic temperature of the triple point of water.

Definitions of SI Base Units

Unit of amount of substance	**mole**	1. The mole is the amount of substance of a system which contains as many elementary entities as there are atoms in 0.012 kilogram of carbon 12; its symbol is "mol." 2. When the mole is used, the elementary entities must be specified and may be atoms, molecules, ions, electrons, other particles, or specified groups of such particles.
Unit of luminous intensity	**candela**	The candela is the luminous intensity, in a given direction, of a source that emits monochromatic radiation of frequency 540×10^{12} hertz and that has a radiant intensity in that direction of 1/683 watt per steradian.

Units Derived from SI Base Units

Derived quantity	Name	Symbol	Expression in terms of other SI units	Expression in terms of SI base units
area	square meter	m^2		m·m
volume	cubic meter	m^3		m·m·m
plane angle	radian	rad		$m\ m^{-1}$
solid angle	steradian	sr		$m^2\ m^{-2}$
speed	meter per second	m/s		m/s
acceleration	meter per second squared	m/s^2		m/s^2
frequency	hertz	Hz		s^{-1}
force	newton	N		$kg{\cdot}m\ s^{-2}$
pressure	pascal	Pa	N/m^2	$kg\ m^{-1}{\cdot}s^{-2}$
energy, work	joule	J	N·m	$kg\ m^2{\cdot}s^{-2}$
power	watt	W	J/s	$kg\ m^2{\cdot}s^{-3}$
electrical charge	coulomb	C		A s
electrical potential, electromotive force	volt	V	W/A	$m^2{\cdot}kg{\cdot}s^{-3}{\cdot}A^{-1}$
electric field strength	volt per meter	V/m	V/m	$m{\cdot}kg{\cdot}s^{-3}{\cdot}A^{-1}$
capacitance	farad	F	C/V	$m^{-2}{\cdot}kg^{-1}{\cdot}s^4{\cdot}A^2$
electrical resistance	ohm	Ω	V/A	$m^2{\cdot}kg{\cdot}s^{-3}{\cdot}A^{-2}$
activity of a radionuclide	becquerel	Bq		s^{-1}
absorbed radiation dose	gray	Gy	J/kg	$m^2{\cdot}s^{-2}$
biological equivalent dose	sievert	Sv	J/kg	$m^2{\cdot}s^{-2}$

Other Units, Abbreviations, and Equivalence Statements

D.M. Hanson

Length		
Name	**Symbol**	**Equivalence Statement**
ångstrom	Å	Å = 0.1 nm = 10^{-10} m
inch	in	1 in = 2.54 cm
foot	ft	1 ft = 12 in
mile	mi	1 mi = 1.609344 km 1 mi = 5280 ft

Mass		
Name	**Symbol**	**Equivalence Statement**
atomic mass unit	u or amu	1 u = $1.6605388 \times 10^{-27}$ kg
pound	lb	1 lb = 0.45359237 kg
metric ton or tonne	t	1 t = 1000 kg
ton	T	1 T = 2000 lb

Volume		
Name	**Symbol**	**Equivalence Statement**
liter	L	1 L = 10^{-3} m^3
gallon	gal	1 gal = 3.78541 L
quart	qt	4 qt = 1 gal
pint	pt	2 pt = 1 qt
ounce	oz	16 oz = 1 pt
tablespoon	T or tbsp	2 tbsp = 1 oz
teaspoon	t or tsp	3 tsp = 1 tbsp

Pressure		
Name	**Symbol**	**Equivalence Statement**
bar	bar	1 bar = 10^5 Pa
atmosphere	atm	1 atm = 101.325 kPa 1 atm = 1.01325 bar
torr	torr	760 torr = 1 atm
pounds per square inch	psi or lb/in^2	14.696 psi = 1 atm

Energy		
Name	**Symbol**	**Equivalence Statement**
electron volt	eV	1 eV = 1.60218×10^{-19} J
calorie	cal	1 cal = 4.184 J
nutritional calorie	Cal	1 Cal = 1000 cal = 1 kcal

Other Units, Abbreviations, and Equivalence Statements (con't)

Temperature		
Name	**Symbol**	**Equivalence Statement**
Celsius temperature	°C	$T_{°C} = T_K$(1 °C / 1 K) – 273.16 °C
Fahrenheit temperature	°F	$T_{°F} = T_{°C}$(180 °F / 100 °C) + 32 °F
Time		
Name	**Symbol**	**Equivalence Statement**
minute	min	1 min = 60 s
hour	h	1 h = 60 min
day	d	1 d = 24 h
Angle		
Name	**Symbol**	**Equivalence Statement**
degree	°	360° = 2π rad
minute	′	60′ = 1°
second	″	60″ = 1′

Physical Constants

http://nist.gov/cuu/Constants/index.html

Quantity	Value*
speed of light in a vacuum (*c*)	2.99792458×10^{8} m/s
elementary charge (+e for proton, –e for electron)	$1.6021765 \times 10^{-19}$ C
electron rest mass (m_e)	$9.1093822 \times 10^{-31}$ kg
proton rest mass (m_p)	$1.67262164 \times 10^{-27}$ kg
neutron rest mass (m_n)	$1.67492721 \times 10^{-27}$ kg
Planck constant (h)	$6.6260690 \times 10^{-34}$ Js
Avogadro constant (N_A or *L*)	6.0221418×10^{23} /mol
Faraday constant (**F**)	96,485.340 C/mol
gas constant (R)	8.31447 J K^{-1} mol^{-1} 0.082057 L atm mol^{-1} K^{-1}
Boltzman constant (k)	1.380650×10^{-23} J/K

* *Values are rounded to include only one uncertain digit.*

Appendix B

Secrets for Success in General Chemistry

Students often ask for advice on how to do better in General Chemistry courses. To provide a response, undergraduates who completed these courses with A and A- grades were interviewed to learn how they did it. ***Here are their secrets, revealed!***

In general:

- Plan to devote 9-12 hours each week to studying chemistry, make a schedule of study times, otherwise you will never put in this much time each week. It just isn't that much fun!
- Be sure to get enough sleep, exercise, and good nutrition always, but particularly before exams.
- Understand the material from each class before the next class.
- If you get stuck, use your textbook as a resource, get help from your friends and classmates, or go to your instructor.
- Study for exams by identifying the key ideas, understanding what they mean, and seeing how they are used in solving problems. Review all the homework problems and connect them to the key ideas.
- Do not simply read and reread the textbook or class notes, rather use these resources to answer specific questions that develop as you do the assigned homework.

Before class:

- Spend a few minutes looking over the reading assignment to identify key points because you learn by building on what you already know.

During class:

- Listen carefully, take notes, and mark issues that are not clear.
- Work with others to complete and understand the Foundations of Chemistry activity and in-class problems.

After class:

- Review your notes, use your textbook (or the instructor's notes if available) to complete them and clarify issues.
- Make sure you understand the in-class activity and sample problems.
- Carefully complete the reading assignment. Take notes to highlight the important points.
- Test your understanding by completing the homework assignment. Use your textbook as a resource to resolve any issues that arise. Anything that you are asked to do is explained in class or in the reading assignment. You just need to understand and put the ideas together.

After you have completed an assignment:

- If your instructor does not score your homework, check your answers and solutions against answers or solutions that are posted, and grade yourself.
- Do not use posted solutions as examples that you read and attempt to memorize. The understanding that you need for success on exams is developed by figuring things out before the solutions are posted, not by reading and memorizing the solutions that someone else prepared.
- If you don't understand a solution, analyze similar examples in your textbook, get help from friends and classmates, or see the instructor.

APPENDIX C

How Your Brain Works

Implications for Learning Chemistry Efficiently

Your brain has a short-term memory or working memory that processes incoming and outgoing information. The capacity of this working memory is small; it can only deal with five to nine pieces of information. The working memory can be expanded by using paper to take notes and work out solutions to problems. With ideas and notes on paper, many items and the connections between them can be visualized.[1]

Your brain has a perception filter that is controlled by prior knowledge and experiences that are stored in long-term memory.[1] The perception filter restricts the sensory information that reaches the working memory. The information that does reach working memory subsequently is stored in long-term memory if properly reinforced by repetition. Since learning depends on what you already know, you need to go over material repetitively, learning a little bit more each time. If you don't understand something, come back to it the next day because your brain continues to work and make connections while you sleep. Do not try to cover too much material in too short of a time because the rate at which you learn is limited by the perception filter, the capacity of working memory, and the rate your brain processes new information. It needs time to consolidate and connect ideas, and a good night's sleep helps. Spread the time you devote to learning chemistry over the entire week; don't concentrate it in a few days.

Your brain has a librarian function. There is no way that words or actions of a teacher can be stored directly in long-term memory and automatically be understood and utilized by the learner. There must be a conversion of this external language into your knowledge system with added connections so the knowledge can be retrieved and used when needed.[2,3] It is the librarian that executes this task. Initially the librarian stores information as disconnected pieces. To make use of this knowledge, the facts, vocabulary, equations, procedures, concepts, and contexts need to be interconnected in hierarchical structures. The following four actions on your part are most important to assist the librarian in making these connections and building these structures.[4]

- Identify the important concepts covered in an assignment and the connections between those concepts.
- Think about a concept in multiple representations: as a verbal statement, using symbols or an equation, and as a picture, diagram, or graph.
- Reflect on the concepts and procedures used in textbook examples or in solving a homework problem or answering a question: what concepts were needed and why were they needed, what were the steps in the procedures and why were those steps necessary, and how are the concepts related to the procedures?
- Solve many different problems with different contexts because repetition is essential. Analyze these different situations to identify similarities and differences in order to build a library of contexts to draw upon by analogy.

Your brain needs a reflector/analyzer. The only way to improve performance is to think about what you have done well, what needs to be improved, what has been learned, and what is not yet understood. Such reflection and analysis is essential for continual improvement, growth, and success.[5]

1 Johnstone, A. H. *J. Chem. Ed.* **1997**, *74*, 262-268.
2 Newell, A.; Simon, H. A. *Human Problem Solving*; Prentice-Hall: Englewood Cliffs, NJ, 1972.
3 Simon, H. A. In *Problem Solving and Education: Issues in Teaching and Research*; Tuma, D. T., Reif, F., Eds.; Lawrence Erlbaum Associates: Hillsdale, NJ, 1980, p 81-96.
4 Hanson, D. M. In *Process-Oriented Guided-Inquiry Learning*; Moog, R. S., Spencer, J. N., Eds.; American Chemical Society: Washington, DC, 2008, p 14-25.
5 *How People Learn: Brain, Mind, Experience, and School*; Bransford, J. D.; Brown, A. L.; Cocking, R. R., Eds.; National Academy Press: Washington, D,C, 2000.